Analysis of a Time-in-Grade Pay Table for Military Personnel and Policy Alternatives

BETH J. ASCH, MICHAEL G. MATTOCK, PATRICIA K. TONG

Prepared for the 13th Quadrennial Review of Military Compensation
Approved for public release; distribution unlimited

NATIONAL DEFENSE RESEARCH INSTITUTE

For more information on this publication, visit www.rand.org/t/RRA369-1

Library of Congress Cataloging-in-Publication Data is available for this publication.
ISBN: 978-1-9774-0583-8

Published by the RAND Corporation, Santa Monica, Calif.
© Copyright 2020 RAND Corporation
RAND® is a registered trademark.

Cover: Military: MTMCOINS/Getty Images
Cover design: Rick Penn-Kraus

Support RAND
Make a tax-deductible charitable contribution at
www.rand.org/giving/contribute

www.rand.org

Preface

Federal law mandates that every four years the Secretary of Defense conduct an assessment of the military compensation system, resulting in a Quadrennial Review of Military Compensation (QRMC). In response to a request articulated in Section 603 of the Senate Armed Services Committee version of the National Defense Authorization Act of 2019, the 13th QRMC is providing an assessment of the effects of a time-in-grade pay table for military personnel, particularly on readiness. A time-in-grade pay table would set pay based on pay grade and years of service within a grade, in contrast to the current time-in-service pay table, which sets pay based on pay grade and years of service in the military. While interest in a time-in-grade pay table is not new, and in fact it was assessed by past commissions, including the 10th QRMC, interest in it has been renewed because of efforts at the congressional level and within the services to more flexibly manage military personnel to attract, retain, and promote better performers. The primary means by which military personnel are financially rewarded for superior performance is through faster promotion, so a time-in-grade pay table may increase performance by providing a permanent reward to those who are promoted faster. The current time-in-service pay table provides only temporary financial rewards to those who are promoted faster.

The 13th QRMC asked the RAND Corporation to assist in its assessment of a time-in-grade pay table. This report describes the results of these analyses. It should be of interest to those concerned about the setting of military pay and its effects on readiness.

The research was sponsored by the 13th QRMC and conducted within the Forces and Resources Policy Center of the RAND National Security Research Division (NSRD), which operates the RAND National Defense Research Institute (NDRI), a federally funded research and development center sponsored by the Office of the Secretary of Defense, the Joint Staff, the Unified Combatant Commands, the Navy, the Marine Corps, the defense agencies, and the defense intelligence enterprise.

For more information on the RAND Forces and Resources Policy Center, see http://www.rand.org/nsrd/frp or contact the center director (contact information is provided on the webpage).

Contents

Figures

Tables

Summary

The U.S. Department of Defense's (DoD's) Thirteenth Quadrennial Review of Military Compensation (13th QRMC) was mandated by Congress to provide an assessment of a time-in-grade (TIG) basic pay table as a replacement for the current time-in-service (TIS) basic pay table. This report summarizes analysis conducted in support of this requirement. Observers as well as past commissions have argued that a TIG pay table would provide stronger incentives for superior performance and better facilitate the lateral entry of personnel with civilian acquired skills, two outcomes that would align with the services' and Congress's objective of improving military personnel talent management.

Each cell of the current TIS basic pay table indicates members' pay based on their pay grade and years of service (YOS) or longevity in the military. Under a TIG pay table, basic pay would be based on pay grade and years in that grade. Given that faster promotion is the primary means by which the services financially reward superior performance, a disadvantage of a TIS pay table is that the financial reward to faster promotion is temporary and only lasts until the rest of the member's cohort is promoted as well. In contrast, a TIG pay table would provide a permanent financial reward to faster promotion, and past studies and commissions have argued that a TIG would thereby increase military personnel performance. Another advantage of a TIG pay table, as argued by its proponents, is that it could improve the pay and therefore the competitiveness of the military to lateral entrants relative to the current TIS pay table.

Research Questions and Approach

To support the 13th QRMC, we developed a TIG pay table building and extending the TIG pay table developed by the 10th QRMC and 2006 Defense Advisory Committee on Military Compensation (DACMC). As with these earlier studies, the TIG pay table we developed sought to keep basic pay over a military career unchanged for those who receive due-course promotions and experience average promotion times. Given this TIG pay table, we then sought to address the following research questions:

1. How would the TIG pay table affect military pay over a military career?
2. Would the TIG pay table better facilitate lateral entry?
3. How would the TIG pay table affect retention, personnel performance, and cost?
4. What would be the cost to DoD to transition to the TIG pay table if DoD sought to hold member harmless in terms of experiencing no pay reductions in the first year of the transition?

5. Can the advantages of the TIG pay table be achieved by reforming the current TIS pay table and/or adopting other pay or personnel policies?

Our approach involved building on and extending the work of past studies and commissions and making use of more-recent data and modeling capabilities, such as RAND's dynamic retention model (DRM). First, we computed how basic pay would change over the course of a career for enlisted members and commissioned officers for those promoted faster or slower than those receiving due-course promotions, as well as for lateral entrants, warrant officers, and officers with prior enlisted service.

Second, we extended the mathematical structure of the DRM to include promotion and estimated the model for enlisted personnel and officers in each service. We then developed DRM simulation capability to allow the simulation of the retention, cost, and performance effects of the TIG pay table we developed versus the TIS pay table. Our DRM simulations assume that promotion speed depends on performance, which in turn, depends on innate ability and effort. We do not observe ability or effort. Instead, we treat effort and ability as unitless indices, and then we make assumptions about how ability and effort affect promotion speed. We also make assumptions about the distribution of ability among entrants, how ability affects external opportunities, and the disutility of increased effort. These assumed parameters are calibrated or chosen so that we can replicate the observed retention profile of enlisted members and officers within each service. We also conducted sensitivity analyses and found that our main conclusions, discussed below, were unchanged qualitatively under alternative assumptions. Because we were more successful in incorporating innate ability than effort into the model, our reporting of results focuses on ability. To report results on ability, we first compute each member's simulated percentile in the ability distribution (e.g., the 50th percentile would represent the mean) and then report the overall ability of the force in terms of the mean ability percentile. To assess the extent to which the TIG improves the selective retention of higher-ability personnel in higher grades, we also report the average ability percentile of higher-grade versus lower-grade personnel.

Third, we estimate the extent to which members' basic pay would increase or decrease in the transition to the TIG pay table and the cost to DoD of providing "save pay" so as to hold members harmless in the first year of the transition. Fourth, we examine whether the advantages of the TIG pay table could be fully achieved by retaining the current TIS pay table and adopting two alternative policies: (1) constructive credit for performance, which would give service members who are promoted faster than their peers a permanent one year of service increment in the pay table for the purpose of computing basic pay, and (2) credential pay, a pay based on skills, knowledge, education, or training credentials. Finally, because critics of the TIG pay table have argued that it would create inequitable pay differences owing to differences in promotion speed unrelated to performance but related to differences in promotion opportunity (supply and demand factors), we investigate the extent to which the TIG pay table provides increased incentives for performance, even after accounting for differences in promotion opportunity.

Findings and Conclusions Regarding the Advantages and Disadvantages of the Time-in-Grade Pay Table

Consistent with the findings of past commissions, we find that the TIG pay table that we developed would provide a permanent financial reward for early promotion, thereby providing greater incentives for performance for both enlisted personnel and commissioned officers. In simulations of basic pay for enlisted personnel, we find that the discounted present value of basic pay is 11.3 percent rather than 5.5 percent higher for those promoted earlier under the TIG versus the TIS pay table and that the discounted present value of retired pay is 22.8 percent rather than 14.3 percent higher. Furthermore, the pay advantage of the TIG pay table for those promoted faster remains, even when we control for factors unrelated to performance, such as supply and demand factors that can alter promotion opportunities at a point in time. Also consistent with past commission findings, a second advantage of the TIG pay table is that it provides higher entry pay than the TIS pay table to lateral entrants.

Unlike past commissions, we also provide estimates of the retention, cost, and performance effects of the TIG pay table. DRM simulations indicate that the TIG pay table would be a more efficient approach to setting basic pay. We show simulation results in Table S.1 using the Army enlisted force as an example. Results for the other services are qualitatively similar.[1]

We find that the average ability percentile across the force increases under the TIG pay table from 47.3 to 48.9. Furthermore, ability sorting improves under the TIG pay table, meaning the TIG pay table is more successful at inducing higher-ability personnel to stay and seek advancement to the upper grades. In particular, under the TIS pay table, the average ability

Table S.1
Army Enlisted Summary Statistics of Retention, Performance, and Cost

Army Enlisted Personnel	TIS Pay Table	TIG Pay Table	TIG Pay Table with 0.375% Across-the-Board Pay Cut
Average ability percentile			
E-5	42.8	43.6	43.7
E-9	66.0	76.9	76.8
Overall	47.3	48.9	48.9
Retention: percentage change in force size	0.0	1.5	0.0
Cost per members (2019 dollars)	$64,324	$64,173	$63,634

SOURCE: Authors' computations.
NOTES: The table shows simulated effects on ability, retention, and cost using the DRM parameter estimates for Army enlisted personnel. The first column shows simulations under the current TIS pay table, the second shows simulations under a proposed TIG pay table and the third shows, for demonstration purposes, the effects of an across-the-board pay reduction that would achieve the same overall retention under the TIG as the TIS pay table. Ability is a unitless measure with which we calibrate the parameters of the distribution of the ability distribution. The table shows the average percentile of the distribution for the force overall and at the grades of E-5 and E-8. Costs include active duty basic pay and allowances and retirement accrual costs.

---

[1]   Our analysis covers enlisted personnel and commissioned officers in the Army, Navy, Marine Corps, and Air Force. We exclude U.S. Coast Guard personnel because the Defense Manpower Data Center data we used exclude these personnel.

percentile of an E-9 is 66.0 compared with 42.8 for an E-5, an increase of 54.2 percent. This effect is stronger under the TIG pay table; the average ability percentile increases 76.3 percent (from 43.6 to 76.9). This result occurs because better performers are more likely to be promoted and retained under the TIG pay table. We find similar results for enlisted personnel in the other services. The table also shows that retention improves by 1.5 percent—the higher retention of better performance more than offsets the lower retention of poorer performers. Although the force becomes larger, cost per member decreases from $64,324 to $64,173.

To show the increased efficiency of the TIG pay table, we simulate, for demonstration purposes, the effects of an across-the-board pay reduction that would achieve about the same overall retention under the TIG as the TIS pay table. We find that a 0.375 percent pay cut would achieve a force of the same size. The key result is that personnel cost per member is even lower, $63,634 versus $64,324, while average ability and ability sorting are improved. Stated differently, the TIG pay table is more efficient because it can achieve about the same retention as the TIS pay table, at less cost per member, and with improved performance.

Another advantage of the TIG pay table is that it can provide stronger retention incentives for occupations and career fields that experience shortfalls as a result of demand and supply factors. For example, when the economy improves and retention falls, promotion opportunities improve in occupations that experience the greatest shortfalls. The improved promotion opportunities act as a self-correcting mechanism by inducing higher retention (or lessening the impact of declining retention) and attracting more personnel to occupations experiencing retention issues. Because the TIG magnifies the financial effects of differences in promotion speed, this self-correcting mechanism is stronger under a TIG pay table. As we discuss below in the context of the disadvantages of the TIG pay table, much but not all of the difference in promotion speed is attributable to these supply and demand factors.

Disadvantages of the Time-in-Grade Pay Table

The TIG pay table is not without disadvantages. The major disadvantage is that the transition would be costly to DoD and would be disruptive to a significant fraction of the force. We estimate that about one-third of the active force (32.1 percent) would experience a basic pay reduction in the transition to the TIG pay table, with an average reduction in basic pay of 6 percent among those who would experience a pay reduction. If DoD were to adopt save pay to hold members harmless, we estimate that, in the first year, the cost would be $1.39 billion, in 2018 dollars. To put this figure in context, the 2018 appropriation for active component military personnel was about $115.9 billion (DoD, 2019).[2] The $1.39 billion figure does not include the cost of providing financial education to the force and "socializing" the change to smooth the transition.

Another challenge with establishing the TIG pay table is that pay for warrant officers and commissioned officers who transition out of the enlisted force could decrease, creating a pay inversion for these personnel. The inversion arises because members promoted from the enlisted force to either the warrant officer or commissioned officer force often have widely different amounts of prior enlisted service. Another reason for the inversion is that the TIG pay table for warrant officers is designed for those without prior enlisted service, so pay decreases for those who become warrant officers with prior enlisted service. This disadvantage of the

---

[2]   This figure excludes Medicare-Retiree Health Care Contributions.

TIG pay table could be addressed by allowing the services to flexibly set the starting grade for those with prior enlisted service. For example, allowing warrant officers with prior enlisted service to transition to warrant officer status at the grade of W-2 or W-3 could address the pay inversion.

Another disadvantage of a TIG pay table noted in the past is that differences in promotion speed can reflect factors other than differences in individual performance, such as differences in promotion opportunities due to supply and demand factors. For example, if the economy improves, retention falls, thereby increasing promotion opportunities for those in the lower grades. We find evidence to indicate that a relatively large share of the variation in promotion is attributable to factors such as supply and demand factors that are unrelated to merit. Further, the TIG pay table would exacerbate the pay differences that result from the variation in promotion. But these other factors do not explain all of the differences in promotion speed. To the extent that the remaining differences in pay, after controlling for these other factors, represent the financial incentive for performance, we find that the remaining differences are still larger under the TIG than the TIS pay table. The implication is that while the criticism has merit, it still the case that the TIG pay table provides a stronger financial incentive for performance.

Could the Advantages of the Time-in-Grade Pay Table Be Fully Achieved with a Time-in-Service Pay Table?

The answer to this question is yes for some advantages of the TIG pay table, but in terms of the major advantages of the TIG pay table—the increased efficiency and performance of the force—the answer is no, though with some changes in policy, a TIS pay table might be able to come close.

An advantage of the TIG pay table is that it would allow pay to be more competitively set for lateral entrants. We find that an identical result could be achieved under a TIS pay table, if Congress changed the current definition of constructive credit to give the services the opportunity to offer not just a higher entry grade but also a higher longevity entry point. For example, a lateral entrant could be permitted to enter as an O-3 with 10 YOS.

We also find that redefining constructive credit to provide YOS credit for performance is a policy that can broadly replicate the higher basic pay found under the TIG pay table. Our DRM simulations indicate that constructive credit for performance would also be an improvement over the TIS pay table (in the absence of constructive credit for performance) in terms of efficiency, at least in terms of ability sorting. But enlisted and officer retention, average ability, and ability sorting would not improve as much as predicted under the TIG pay table. In other words, the simulations indicate that constructive credit is an improvement over the current TIS pay table but would be less efficient than the TIG pay table.

We also examined whether credential pay is a policy that could provide performance incentives under a TIS pay table and found that credential pay is not designed to be a pay-for-performance program that rewards superior performance and reduces pay for those who fall short. Thus, it would not be an effective substitute to the TIG pay table in terms of increasing performance incentives.

Closing Thoughts

The TIG pay table would better support service and congressional efforts to improve talent management. But transitioning to the TIG pay table would involve costs, not the least of which is the disruption to the force regarding a fundamental feature of their service—namely, how they are paid. Although constructive credit for performance could achieve some of the advantages of the TIG pay table, simulations suggest that it would not be quite as efficient or performance-enhancing as the TIG pay table. One approach to implementing the TIG pay table while minimizing risk is to do so on an experimental basis as the TIG demonstration project. Doing so would enable DoD to fully assess the full array of transition costs, permit the development of effective financial education, and allow further assessment of the retention, cost, and performance effects of the TIG pay table.

Acknowledgments

We would like to thank Thomas Emswiler and Colonel Brunilda Garcia, Director and Associate Director, respectively, of the 13th QRMC for their guidance and input on this project. At RAND, we wish to thank Hannah Han and Tony Lawrence. We are also grateful to the reviewers of this document, Michael Hansen at RAND and Patrick Mackin of SAG Corporation.

Abbreviations

ADSO	active duty service obligation
DACMC	Defense Advisory Committee on Military Compensation
DLPT	Defense Language Proficiency Test
DMDC	Defense Manpower Data Center
DoD	U.S. Department of Defense
DOPMA	Defense Officer Personnel Management Act
DPV	discounted present value
DRM	dynamic retention model
FLPB	Foreign Language Proficiency Bonus
NDAA	National Defense Authorization Act
PCMC	President's Commission on Military Compensation
QRMC	Quadrennial Review of Military Compensation
RMC	Regular Military Compensation
TIG	time in grade
TIS	time in service
WEX	work experience file
YOS	years of service

CHAPTER ONE

Introduction

Section 603 of the Senate Armed Services Committee (SASC) version of the National Defense Authorization Act (NDAA) of 2019 requested that the U.S. Department of Defense (DoD) submit a report on setting a time-in-grade (TIG) pay table for military personnel, as a replacement for the current time-in-service (TIS) basic pay table. It also requested an assessment of the effects of a TIG pay table on readiness. Every four years, DoD conducts a Quadrennial Review of Military Compensation (QRMC). Because the work of the 13th QRMC was just being launched at the time of the Senate Armed Services Committee request, the analysis to support the DoD response to the congressional request was folded into the mandate of the 13th QRMC. The 13th QRMC requested that RAND support its effort and provide analyses of a TIG pay system for military personnel.

Basic pay is the foundation of military compensation, making up about 60 percent of Regular Military Compensation (RMC), the military's rough equivalent of a civilian salary (Asch, Hosek, and Martin, 2002).[1] Every service member on active duty is entitled to basic pay, given by a TIS pay table in which the amount of pay depends on the member's pay grade and length of service. The structure of the current pay table was created just after World War II, and while the pay table has changed over time—for example enlisted pay grades were added in 1958, and the pay table was extended to 40 years of service (YOS) in 2007—the table's basic structure, and the fact that pay is based on rank and YOS, has remained unchanged.

An alternative approach to setting the pay table is to base the amount of pay on rank and steps in grade within a grade, otherwise known as time in grade. The federal general schedule pay table is an example of a TIG system. The pay of federal employees is based on their grade, e.g., GS-9, and their pay step within a grade. Importantly, years of experience is not used for computing the amount of pay. For military personnel, a TIG pay table would base monthly basic pay on rank and years served within a given grade.

The issue of whether a TIG pay table is preferred over a TIS pay table is related to the question of whether military's promotion system sufficiently rewards personnel who perform better and whether the promotion system embeds strong enough incentives for performance. The services primarily reward performance through the promotion system, whereby those who have demonstrated superior performance are rewarded by being promoted faster than their peers. When service members are promoted earlier than their peers, they move up a grade and receive the higher pay associated with that grade, thereby experiencing a pay advantage over those who receive a due-course promotion. Under a TIS pay table, that pay advantage disap-

---

[1] RMC consists of basic pay, the basic allowance for housing, the basic allowance for subsistence, and the tax advantage from receiving allowances tax-free.

pears once the peers receive their promotion. For example, an E-5 who is promoted to E-6 at 5 YOS rather than at 6 YOS receives the pay of an E-6 with 5 YOS. But, once the early promotee's peers are promoted at 6 YOS, both the early promotee and their peers receive the same pay, namely that of an E-6 with 6 YOS.

In contrast, under a TIG pay table, the pay advantage of the early promotee would be permanent. Thus, the financial reward for better performance and faster promotion would be greater under a TIG table. A TIG table would also allow the services to offer higher pay to lateral entrants, meaning individuals with relevant civilian experience who enter the military at a higher pay grade. Under a TIS pay table, a lateral entrant would enter at a higher grade but at the lowest YOS cell in the pay table. Under a TIG pay table, a service member's pay is not constrained by their lack of YOS.

While interest in a TIG table for military personnel is not new, interest in it has been renewed because of efforts at the congressional level and within the services to more flexibly manage military personnel to attract, retain, and promote better performers. The 2019 NDAA included reforms to the 1980 Defense Officer Personnel Management Act (DOPMA) that authorize the services to grant "constructive credit" for education and for work experience, thereby allowing individuals to enter service at a rank as high as an O-6 (colonel or Navy captain). The reforms also allow the services to suspend "up-or-out" requirements for some types of officers so that officers are granted more opportunities for promotion to a higher grade. The 2019 NDAA also allows better-performing officers to be placed higher on promotion lists than their peers, changing the traditional seniority-based system.

Each of the services is also focusing on improved talent management. For example, the Army created the Army Talent Management Task Force. Among its initiatives is a pilot program that allows officers and units to participate in a marketplace and submit preferences for each other. It also includes brevet or temporary officer promotions for critical shortage areas, as well as promotions for enlisted noncommissioned officers that are based on performance and not just on their seniority and rank relative to peers. While these efforts and legislative changes focus on personnel management rather than compensation, the adequacy of military compensation in supporting these efforts must also be considered. In particular, as stated in Senate Armed Services Committee testimony by former Under Secretary of Defense for Personnel and Readiness David Chu, a TIG pay approach might better support new authorities granted by Congress (Chu, 2018).

Research Questions and Approach

To support the 13th QRMC's assessment of the TIG pay table approach, we developed a TIG pay table, building on past studies and commissions, as described below. Given this TIG table, we then sought to address the following research questions:

1. How would the TIG pay table affect military pay over a career?
2. Would the TIG pay table better facilitate lateral entry of personnel with relevant civilian experience?
3. How would the TIG pay table affect retention, personnel performance, and cost?
4. How much would it cost DoD to transition to the TIG pay table if DoD sought to hold members harmless in terms of experiencing no pay reductions in the first year of the transition?

5. Can the advantages of the TIG pay table be achieved by reforming the current TIS pay table and/or adopting other pay or personnel policies?

To address the first two questions, we used the TIG pay table we developed and assessed how pay would change over the career of enlisted members and commissioned officers for those promoted faster or slower than those receiving due-course promotions. We also computed pay over a career for warrant officers with no prior service, for enlisted members who transition to warrant or commissioned officer status, and lateral entrants.

Addressing the third question requires a modeling capability that can address "what if" questions about how retention and cost would change under as-yet-untried changes to the structure of military compensation. The capability needs to be based on a solid theory of retention decisionmaking over a service member's career, empirically grounded in data on actual retention behavior of service members over a long period of time, and it needs a simulation capability that allows us to assess major compensation reforms without relying on the existence of prior experience with such reforms. RAND's dynamic retention model (DRM) provides such a capability. The model is a stochastic dynamic programming model of the individual's decision to stay or leave active duty and, if a member leaves, the decision to participate or not in the reserves. The model has several rich and realistic features. It's a lifecycle model in which retention decisions are based on forward-looking behavior that depends on current and future military and civilian compensation. It allows for uncertainty in future periods and recognizes that people may change their mind in the future as they get more information about the military and their external opportunities. It also recognizes that individuals differ in their preferences for service in the active or reserve components. Furthermore, the model is formulated in terms of the parameters that underlie the retention and reserve participation decision processes rather than on the average response of the population of members to a particular compensation policy. As a structural model, it is well suited to permit assessments of alternative compensation systems that have yet to be tried.

To address the third question, we extended RAND's DRM capability to permit assessment of the retention, cost, and performance effects of the TIG table versus the TIS pay table. This task required that we extend the mathematical structure of the model and develop appropriate computer code to incorporate grade in the estimation and simulation capabilities for enlisted personnel and officers in each service. We estimate the model using longitudinal data on individual service members in each service that track their careers from entry, as far back as 1990, to the present. Once estimated, we then used the model estimates to simulate the retention, cost, and performance implications of the TIG pay table.

We address the fourth question by using Defense Manpower Data Center (DMDC) data on current time in grade and time in service for all active duty military personnel as of January 2019 to assess the extent to which pay would be lower for personnel during a transition to the TIG pay table and the cost of restoring pay to pre-transition levels for a year.

For the fifth question, we use the DRM capability to simulate alternative pay and personnel policies that might be implemented under a TIS pay table, such as constructive credit for performance. In addition, we review the literature on the feasibility of using credential pay—a pay based on skills, knowledge, education, or training credentials—to increase performance incentives under the current TIS pay table.

Our approach builds on and extends past analyses of the feasibility and desirability of the TIG table. To better highlight the ways in which we extend these past efforts, we first briefly review the findings of previous commissions and study groups. A summary of these past efforts in provided in Table 1.1.

Previous Commission and Study Group Findings[2]

The Hook Commission developed the modern-day TIS pay table in 1948. In doing so, it considered the performance incentives associated with longevity increases and the appropriate structure for embedding these incentives. In particular:

> Increases for length of service should provide a stimulus to do better work but should cease after a reasonable period of time so that a lower level of responsibility will not receive the pay of a higher level and thus remove the incentive of striving for promotion. (Advisory Commission on Service Pay, 1948, p. 2)

But, ultimately, the Hook Commission fell short of recommending a TIG table over a TIS one.

The 1957 Defense Advisory Committee on Professional and Technical Compensation agreed with the Hook Commission about the need to properly structure longevity increases. It expressed concern about pay inversions in the pay table whereby the pay of personnel in lower grades exceeded the pay of personnel in higher grades, and stated that the "longevity pay system actually rewards, in many cases, the type of men who have little ambition to achieve higher responsibility" (p. 48). The committee called for a new pay table that would replace longevity increases with within-grade merit step increases, i.e., a type of TIG pay table. The purpose would be to eliminate the pay inversions and to encourage meritorious performance. It also recommended that "save pay" be used in the transition so that members would not see a reduction in pay, but stated that members with many years in a given grade may find "his pay frozen at its present level until he qualifies for promotion" (p. 48). As succinctly put by the committee, "In the future, there should be no additional monetary recognition for the professional laggard."

The first QRMC in 1967 disagreed with the 1957 committee. It found that "[Longevity] is the proper basis for in-grade salary increments . . ." and that in-grade increases should reward the growth in productivity associated with greater experience and "long and faithful service, especially for those who, through no fault of their own, face limited promotion prospects" (p. 79). The QRMC argued that a TIS table is more appropriate for two reasons. First, it concluded that most differences in promotion times reflected differences in promotion opportunity rather than differences in individual merit. This conclusion was based on tabulations that showed that the average time in service at each grade varied across service, rather than analysis that decomposed promotion timing to the portion attributable to promotion opportunity versus individual factors. Second, the first QRMC concluded that the military's "in-at-the-bottom, up-through-the-ranks" personnel management approach meant that experience over a career, rather than within a particular grade, was a more important contributor

---

[2]   Several past studies have reviewed the literature on a TIG versus TIS pay table approach. These include studies by the Congressional Budget Office (1995), the Defense Advisory Committee on Military Compensation (2006), and the 10th QRMC (DoD, 2008).

Table 1.1
Overview of Past Commissions and Study Groups That Have Examined a Time-in-Grade Pay Table

Commission	Report Date	Supported TIG Pay Table?
Hook Commission	1948	No
Defense Advisory Committee on Professional and Technical Compensation	1957	Yes
First QRMC	1967	No
President's Commission on Military Compensation	1978	Yes
7th QRMC	1992	No
Defense Advisory Committee on Military Compensation	2006	Yes
10th QRMC	2008	No

SOURCES: Congressional Budget Office (1995), DACMC (2006), and the 10th QRMC (DoD, 2008).

to an individual's military productivity. Thus, it recommended that the longevity structure be retained as the basis for in-grade salary increases.

The 1978 President's Commission on Military Compensation (PCMC) argued that outstanding performance should receive a greater reward than is provided by the current system and that a TIG pay table offers such recognition without having to fundamentally change the promotion process. The PCMC also stated that increasing pay differentiation for outstanding personnel would also help retain these individuals. The PCMC did not recommend a specific TIG table but provided two guiding design principles. It should

- provide for rapid pay increases during the early years in grade, with a leveling out in later years
- allow for overlap with adjacent pay grades to ensure that the retention of individuals with no promotion potential but nevertheless have value in their current grade.

Like the 1st QRMC, the 7th QRMC in 1992 also rejected the TIG approach. It also raised the question of whether the relevant work experience that is considered in the determination of pay should be an entire military career or time spent in a particular pay grade. It argued that it should be an entire military career. Further, the 7th QRMC found that there are significant differences in promotion timing among skills in the same service and across services, resulting in pay differences among these skills and services. To the extent that these differences are due to supply differences, e.g., retention differences across skills and services, the resulting pay differences may be desirable because they create a self-adjusting pay mechanism to address retention issues. This self-adjusting mechanism works under both a TIS and TIG pay table, though the boost in retention incentives would be larger under a TIG pay table because pay differences associated with promotion would be larger. Nonetheless, these pay differences would be perceived as inequitable because they are not due to differences in performance. The 7th QRMC recommended that adjustments be made to the TIS pay table to offer greater rewards for performance, such as increasing pay raises associated with increases in rank relative to time in service.

The Defense Advisory Committee on Military Compensation (DACMC) was chartered to identify approaches to balance military pay and benefits to sustain the recruitment and retention of high-quality people. Among the topics it considered was pay for performance. Like earlier study groups, the DACMC report in 2006 recommended that performance incentives for early promotion be increased by moving to a TIG pay table. Unlike previous study groups, except for the 1957 Defense Advisory Committee on Professional and Technical Compensation, the DACMC provided an example of a TIG basic pay table, based on the 2005 current TIS table. Furthermore, it showed that under the example, pay differences would be greater over a career for those promoted earlier than for those who received due-course promotions. The DACMC was also the first to recognize that a TIG pay table would be more attractive to individuals with prior service or those who are lateral entrants with specialized skills. The DACMC noted that a TIG pay table could cause pay inversions for enlisted members who transition to warrant officer of commissioned officer, but also noted that the services could transition these members to a higher officer grade. It also discussed the need to ensure that no member sees a nominal decrease in their pay during the transition period from a TIS pay table to a TIG table and explained that this could be avoided through a "save pay" provision. The DACMC estimated that a transitional save pay provision would cost about $1.1 billion (in 2005 dollars).

The 10th QRMC (2008) expanded on the DACMC study, though, unlike the DACMC, the 10th QRMC did not recommend a TIG pay table. The 10th QRMC further developed the TIG pay table example from the DACMC, basing it on the 2007 TIS pay table. Other developments included extending the table through 14 YOS within a grade (versus 9 in the DACMC example), though the 10th QRMC curtailed TIG pay increases at the lower pay grades for both officers and enlisted members. It also addressed potential pay inversions for personnel in grades O-1E to O-3E and for warrant officers.[3] Like the DACMC, the 10th QRMC showed that the pay premium over a career is larger under a TIG table than under a TIS table for those promoted faster, but it also expressed concern about a TIG pay table. As with previous study groups, the 10th QRMC was concerned that promotion speed does not always reflect merit but could reflect supply and demand conditions. The DACMC (2006) also noted this concern but stated that the concern is also relevant for the TIS pay table, so "the criticism is a matter of degree, not kind" (p. 46).

The 10th QRMC also discussed the variation in compensation that currently exists among members entering the warrant officer ranks, making it difficult to devise an entry level pay rate for warrant officers under a TIG pay table. For example, a TIG pay table that sought to maintain the pay of more-senior enlisted personnel who become warrant officers would result in substantial pay raises for warrant officers without military experience. The 10th QRMC also argued that a TIG pay table would result in a major overhaul of the current pay table to improve the compensation of the small percentage of the force that is promoted early. The 10th QRMC dismissed the argument that, although relatively few service members would have a change in compensation under a TIG pay table, the incentive effects of the improved compensation could be force-wide.

An additional concern raised by the 10th QRMC is that a TIG pay table would result in different retired pay amounts for personnel who served the same amount of total service and

---

[3]   The grades O-1E to O-3E are for enlisted members or warrant officers who become commissioned officers. The 10th QRMC addressed the potential pay inversion as these members transitioned from the grade of O-3E to O-4.

achieved the same final grade. A counterargument, one rejected by the 10th QRMC, is that this difference in retired pay is part of the overall performance incentive provided to members who are promoted early and so can be viewed as a desirable feature of the TIG pay system.

While the 10th QRMC did not support adoption of a TIG pay table, it did support the conclusion of the DACMC about the need to embed stronger incentives for performance. For that reason, it made two recommendations.

First, it recommended that the TIS calculation under the current pay table be adjusted through a policy of "constructive credit." Under this policy, the services could credit members with extra YOS, i.e., grant constructive credit, for the purpose of computing their basic pay (but not their retired pay). Fast promotees could be awarded credit for an additional year of service, allowing the member to "move up" a cell within the pay table, relative to peers. Such constructive credit could provide a permanent pay differential to those promoted early. The 10th QRMC argued that this approach would also work for lateral entrants by giving YOS credit to those with prior service or relevant civilian experience. Constructive credit already exists as part of DoD personnel policy, but under current authority lateral entrants may enter a service at a higher grade, but only at the lowest TIS pay cell within that grade. Under the 10th QRMC proposal, entrants could be placed not only at a higher grade but at a higher TIS pay cell for that grade.

Second, the 10th QRMC recommended that the services explore other pay for performance incentives, including credential pay and performance-based bonuses. Credential pay would reward members who receive certification in critical skills. Performance-based bonuses could be a new type of special and incentive pay. Alternatively, the services could introduce a performance element into existing bonuses, such as tying reenlistment bonuses to performance.

Table 1.2 summarizes the main advantages and disadvantages of a TIG pay table approach identified in these past efforts as well as approaches that could be used under the current TIS pay table approach to increase incentives for performance.

How This Study Builds on and Extends Past Efforts

Our project builds on these earlier studies, especially the DACMC and the 10th QRMC. First, to develop a TIG table for the 13th QRMC, we started with the TIG pay table developed by the 10th QRMC and updated it using more recent data on average promotion time to each grade, and we addressed some pay inversions in the 10th QRMC whereby pay declined with grade or with promotion. Second, we considered how pay varies over a career for fast versus due-course promotees under a TIG versus a TIS pay table, not only for enlisted personnel and commissioned officers, but also for lateral entrants, warrant officers, and enlisted personnel who transition to commissioned officer or warrant officer status. Third, while past study groups hypothesized how a TIG pay table would affect retention and performance, no prior study provided estimates of these effects. Our project extends the DRM to provide an empirically based assessment of the retention, performance, and cost effects of a TIG versus a TIS pay table. Also, unlike prior efforts, this study also provides estimates of the retention, performance, and cost effects of alternative policies under a TIS pay table approach that might replicate the advantages of a TIG pay table, such as performance-based bonuses. In short, this study provides additional and new evidence on the effects of a TIG pay table.

Table 1.2
Summary of Commission and Study Group Findings

Advantages of a TIG Pay Table	Disadvantages of a TIG Pay Table	Policies to Implement Under a TIS Pay Table
• Provides permanent financial reward for faster promotion, thereby increasing performance incentives across the force and strengthening the self-correcting retention response of changes in promotion speed owing to supply-and-demand factors • Increases retention incentives for better performers • Better facilitates competitive compensation for lateral entrants	• Results in inequitable pay differences over a career and differences in retired pay for members who have different promotion speeds owing to differences in promotion opportunity (supply and demand factors) and not individual merit • Does not recognize the importance of experience in determining pay in an organization where most members enter at the bottom and rise through the ranks • Does not handle well the pay for warrant officers and could result in pay inversions for enlisted members who become either warrant or commissioned officers • Results in transition costs ("save pay") so as to hold nominal pay constant in the transition	• Constructive credit • Proficiency pay • Performance-based bonuses

SOURCES: Congressional Budget Office (1995), DACMC (2006), and the 10th QRMC (DoD, 2008).

Organization of This Report

This report provides a summary of our analysis and findings. In the next chapter, we discuss how we developed a TIG pay table based on the earlier work of the DACMC and 10th QRMC. We also present computations of pay at promotion and over a career under the TIG versus the TIS pay table. We present our extension of the DRM in Chapter Three. We show the updated mathematical structure, model estimates, and the fits of the models relative to the observed data, and we describe the development of the simulation capability. The simulations involve incorporating parameters related to performance and specifically member's innate ability and their work effort. We discuss how we incorporate these parameters and the assumptions we make based on past studies. In Chapter Four, we show TIG versus TIS simulation results for the steady state and specifically the DRM estimates of how the TIG pay table would affect steady-state retention, performance, and cost for enlisted personnel and commissioned officers in each service. In Chapter Five, we show estimates of the extent to which service members might experience a pay reduction during the transition to the TIG pay table regime and provides estimates of the cost of a "save pay" policy for the first year. In Chapter Six, we consider policies that could be implemented under the current TIS pay table that might replicate the advantages of the TIG pay table. In particular, we show DRM estimates of pay policies to increase performance incentives under the current TIS pay table approach and review the available literature on credential pay. In Chapter Seven, we show the extent to which promotion timing reflects factors other than performance, using recent data on promotion timing across the services. This analysis investigates whether the evidence supports one of the critiques of the TIG table: that the TIG pay table exacerbates the pay differences associated with promotion when promotion is driven mostly by non-performance-related factors, such as supply and demand conditions. In Chapter Eight, we summarize our results and discuss the merits and drawbacks of the TIG pay table in light of the new analysis provided by this project.

CHAPTER TWO

A Time-in-Grade Pay Table and Estimates of Basic Pay over a Career

We developed a TIG pay table for the 13th QRMC, building on the sample table produced for the 10th QRMC.[1] We updated the 10th QRMC TIG table in several ways. First, we based the updated TIG table on the January 2018 basic pay (TIS) table, shown in Table A.1 in Appendix A. Second, like the 10th QRMC, we imputed pay for certain cells in the TIG table to prevent pay decreases or inversions when members are promoted and to ensure that members receive a pay increase over the first five years in a given grade.[2] Third, we used data on average times to promotion for 2013–2018 to develop the updated TIG pay table, a more recent period that the early to mid-2000s used by the 10th QRMC to create its TIG pay table. As shown in Table 2.1, promotion times between 2013 and 2018 differed somewhat from those used by the 10th QRMC.

Table 2.1
Years of Service at Promotion to Grade

Grade	10th QMRC	Average 2013–2018	Grade	10th QMRC	Average 2013–2018	Grade	10th QMRC	Average 2013–2018
E-9	22	22	O-10	34	32	W-5	25	20
E-8	20	18	O-9	30	30	W-4	21	14
E-7	14	13	O-8	30	28	W-3	18	9
E-6	8	8	O-7	26	25	W-2	11	5
E-5	4	4	O-6	20	20	W-1		
E-4	1	2	O-5	16	14			
E-3	0	1	O-4	9	9			
E-2	0	0	O-3	4	3			
E-1			O-2	1	1			
			O-1					

SOURCE: DMDC tabulations.

---

[1] The 10th QRMC analysis was performed and summarized in Hogan and Mackin (2008), and the discussion of the 10th QRMC's analysis discussed in this chapter draws heavily from the Hogan and Mackin report.

[2] The imputations were made by taking the average of pay in the neighboring cells. For example, to impute pay for a member with five years in a given grade, we took the average of pay for those with four years and those with six years. Note that the cells that are imputed for the updated TIG table are not identical to the ones imputed by the 10th QRMC.

Table 2.2 shows the updated TIG pay table built for the 13th QRMC. The cells in which pay was imputed are highlighted in yellow. The first column, called "Entry YOS," shows the YOS in the TIS pay table that defines pay at entry to a given grade. For example, the pay of an E-6 with 0 YOS in the TIG pay table is equivalent to the pay of an E-6 with 6 YOS in the TIS pay table. As shown in Table 2.1, the average time to promotion to E-6 between 2013 and 2018 was 6 years. Because the TIG table was built using the average promotion times that have prevailed under the TIS pay table between 2013 and 2018, by design, basic pay over a career under the TIG pay table is nearly identical to that under the TIS pay table, as shown in Figure 2.1 for enlisted personnel and officers. The use of average promotion times or "due-course" promotions implies that pay over a career is the same under the TIS and TIG pay tables for members receiving due-course promotions. It is important to note that the estimates presented in this chapter and in subsequent chapters are specific to the TIG pay table we developed.

In the remainder of this chapter, we show computations of pay at promotion and over a career under the TIG versus TIS pay tables. Example computations are made for the following groups:

- members with differences in promotion timing
- warrant officers with prior enlisted service
- commissioned officers with prior enlisted service
- lateral entrants.

Effects on Pay of Time in Grade for Members with Differing Promotion Times

The key advantage of a TIG pay table over a TIS one is that it potentially provides a greater financial reward for early promotion and a greater financial disadvantage for later promotion. As discussed in Chapter One, a member who is promoted one year earlier compared with an on-time due-course promotion results in a higher rate of pay that is permanent under a TIG pay table but only for one year under a TIS pay table. Consequently, a TIG pay table provides greater incentives for performance, given that fast promotion is the primary means by which the military rewards better performance.

Figure 2.2 shows simulations of monthly basic pay over a career under the TIS pay table (left panel) versus the TIG pay table (right panel) for enlisted members who are promoted faster, slower, or average to E-5 and E-6. At a given YOS, the difference in basic pay for fast versus due-course or slow promotees reflects the financial reward to faster promotion. Under the TIS pay table (left), those promoted faster (green line) receive higher pay for a year or two but the higher rate is temporary because those promoted on time (blue line) eventually catch up. In contrast, under the TIG, the higher pay rate for fast promotes is permanent, and those receiving due-course promotions do not catch up. Consequently, basic pay over a career is higher for those promoted faster, and lower for those promoted more slowly, under the TIG versus the TIS pay table.

Figure 2.3 shows similar simulations for officers who are promoted early versus on-time to O-4. Officers differ from enlisted personnel in that they are considered for promotion by entry-year group and are either promoted or not promoted at specific YOS points. For example, promotion from O-3 to O-4 usually occurs at around the 10th year. By contrast, enlisted per-

Table 2.2
Proposed Time-in-Grade Monthly Basic Pay Table for January 2018 (0–10 Years in Grade)

Entry YOS	Grade	Years in Grade										
		0	1	2	3	4	5	6	7	8	9	10
Commissioned Officers												
28	O-10	15,800.10	15,800.10	15,800.10	15,800.10	15,800.10	15,800.10	15,800.10	15,,800.10	15,800.10	15,800.10	15,800.10
26	O-9	15,800.10	15,800.10	15,800.10	15800.10	15,800.10	15,800.10	15,800.10	15800.10	15,800.10	15,800.10	15,800.10
24	O-8	14,268.30	14,268.30	14,625.60	14,625.60	14,625.60	14,625.60	14,991.00	14,991.00	14,991.00	14,991.00	14,991.00
22	O-7	12,591.90	12,656.40	12,656.40	12,656.40	12,656.40	12,909.60	12,909.60	12,909.60	12,909.60	12,909.60	12,909.60
19	O-6	10,295.70	10,295.70	10,431.15	10,566.60	10,703.85	10,841.10	11,372.40	11,372.40	11,372.40	11,372.40	11,599.80
14	O-5	8,022.30	8,275.95	8,529.60	8,650.05	8,770.50	8,770.50	9,009.30	9,009.30	9,280.20	9,280.20	9,280.20
9	O-4	6,601.20	7,052.70	7,228.20	7,403.70	7,525.65	7,647.60	7,647.60	7,788.00	77,88.00	7,869.30	7,869.30
3	O-3	5,069.70	5,527.80	5,660.40	5,793.00	5,938.20	6,083.40	6,083.40	6,271.20	6,271.20	6,580.20	6,580.20
1	O-2	3,580.50	4,077.90	4,696.20	4,854.90	4,905.00	4,955.10	4,955.10	4,955.10	4,955.10	4,955.10	4,955.10
0	O-1	3,107.70	3,171.30	3,234.90	3,910.20	3,910.20	3,910.20	3,910.20	3,910.20	3,910.20	3,910.20	3,910.20
Commissioned Officers with over 4 Years of Active Duty Service as an Enlisted Member or Warrant Officer												
10	O-3E	6,271.20	6,315.10	6,359.30	6,403.82	6,435.84	6,451.93	6,468.02	6,484.19	6,500.36	6,516.61	6,532.86
8	O-2E	5,112.60	5,245.80	5,379.00	5,481.90	5,584.80	5,584.80	5,738.10	5,738.10	5,738.10	5,738.10	5,738.10
6	O-1E	4,175.40	4,252.65	4,329.90	4,408.80	4,487.70	4,487.70	4,642.80	4,642.80	4,854.90	4,854.90	4,854.90
Warrant Officers												
20	W-5	7,614.60	7,807.65	8,000.70	8,144.55	8,288.40	8,288.40	8,606.70	8,606.70	8,606.70	8,606.70	9,037.80
14	W-4	6,172.50	6,313.35	6,454.20	6,569.55	6,684.90	6,684.90	6,909.60	6,909.60	7,239.90	7,239.90	7,511.10
9	W-3	4,815.30	5,174.10	5,258.70	5,343.30	5,441.10	5538.90	5,538.90	5739.90	5,739.90	6,102.30	6,102.30
5	W-2	3,957.60	4,182.30	4,356.60	4,530.90	4,617.30	4,703.70	4,703.70	4,,873.80	4,873.80	5082.00	5,082.00
0	W-1	3,037.50	3,201.00	3,364.50	3,452.40	3,638.10	3,638.10	3,857.70	3,857.70	4,181.70	4,181.70	4,332.60
Enlisted Members												
22	E-9	6,306.60	6,306.60	6,431.40	6,556.20	6,747.60	6,939.00	6,939.00	6,939.00	7,285.50	7,285.50	7,285.50
18	E-8	5,099.70	5,168.55	5,237.40	5,354.55	5,471.70	5,471.70	5,601.90	5,601.90	5,921.70	5,921.70	5,921.70
13	E-7	4,186.80	4,368.90	4,431.00	4,493.10	4,559.10	4625.10	4,625.10	4,676.10	4,676.10	4,848.30	4,848.30
6	E-6	3,453.60	3,508.65	3,563.70	3,670.20	3,776.70	3,776.70	3,841.50	3,841.50	3,888.90	3,888.90	3,944.10
3	E-5	2,733.30	2,733.30	2,829.30	2,925.30	3,025.50	3,125.70	3,290.70	3,290.70	3,310.50	3,310.50	3,310.50
2	E-4	2,248.50	2,370.30	2,490.60	2,490.60	2,543.55	2,596.50	2,596.50	2,596.50	2,596.50	2,596.50	2,596.50
1	E-3	1,931.10	2,052.30	2,176.80	2,176.80	2,176.80	2,176.80	2,176.80	2,176.80	2,176.80	2176.80	2,176.80
0	E-2	1,836.30	1,836.30	1,836.30	1,836.30	1,836.30	1,836.30	1,836.30	1,836.30	1,836.30	1,836.30	1,836.30
0	E-1	1,638.30	1,638.30	1,638.30	1,638.30	1,638.30	1,638.30	1,638.30	1,638.30	1,638.30	1,638.30	1,638.30
0	E-1<4	1,514.70	0.00	0.00	0.00	0.00	0.00	0.00	0.00	0.00	0.00	0.00

SOURCE: Authors' calculations.

NOTE: Yellow highlighted cells are values that are not derived from the TIS monthly basic pay table (Table 2.1) but are imputed, as described in the main text.

Table 2.2—continued
Proposed Time-in-Grade Monthly Basic Pay Table for January 2018 (11–20 Years in Grade)

Entry YOS	Grade	Years in Grade									
		11	12	13	14	15	16	17	18	19	20
Commissioned Officers											
28	O-10	15,800.10	15,800.10	15,800.10	15,800.10	15,800.10	15,800.10	15,800.10	15,800.10	15,800.10	15,800.10
26	O-9	15,800.10	15,800.10	15,800.10	15,800.10	15,800.10	15,800.10	15,800.10	15,800.10	15,800.10	15,800.10
24	O-8	14,991.00	14,991.00	14,991.00	14,991.00	14,991.00	14,991.00	14,991.00	14,991.00	14,991.00	14,991.00
22	O-7	12,909.60	12,909.60	12,,909.60	12,909.60	12,909.60	12,909.60	12,909.60	12,909.60	12,909.60	12,909.60
19	O-6	11,599.80	11,599.80	11599.80	11,599.80	11,599.80	11,599.80	11,599.80	11,599.80	11,599.80	11,599.80
14	O-5	9,280.20	9,280.20	9,280.20	9,280.20	9,280.20	9,280.20	9,280.20	9,280.20	9,280.20	9,280.20
9	O-4	7,869.30	7,869.30	7,869.30	7,869.30	7,869.30	7,869.30	7,869.30	7,869.30	7,869.30	7,869.30
3	O-3	6,741.60	6,741.60	6,741.60	6,741.60	6,741.60	6,741.60	6,741.60	6,741.60	6,741.60	6,741.60
1	O-2	4,955.10	4,955.10	4,955.10	4,955.10	4,955.10	4,955.10	4,955.10	4,955.10	4,955.10	4,955.10
0	O-1	3,910.20	3,910.20	3,910.20	3,910.20	3,910.20	3,910.20	3,910.20	3,910.20	3,910.20	3,910.20
Commissioned Officers with over 4 Years of Active Duty Service as an Enlisted Member or Warrant Officer											
10	O-3E	6,532.86	6,532.86	6532.86	6,532.86	6,532.86	6,532.86	6,532.86	6,532.86	6,532.86	6,532.86
8	O-2E	5,738.10	5,738.10	5,,738.10	5,738.10	5738.10	5,738.10	5738.10	5,738.10	5,738.10	5,738.10
6	O-1E	4,854.90	4854.90	4,854.90	4,854.90	4854.90	4,854.90	4,,854.90	4,854.90	4,854.90	4,854.90
Warrant Officers											
20	W-5	9,037.80	9,037.80	9,037.80	9,489.00	9,489.00	9,489.00	9489.00	9,964.20	9,964.20	9,964.20
14	W-4	7,511.10	7,820.70	7,820.70	7,820.70	7,820.70	7,976.70	7,976.70	7,976.70	7,976.70	7,976.70
9	W-3	6,346.80	6,346.80	6,492.90	6,492.90	6,648.30	6,648.30	6,860.10	6,860.10	6,860.10	6,860.10
5	W-2	5,244.60	5,244.60	5,391.90	5,391.90	5,568.30	5,568.30	5,684.10	5,684.10	5,775.90	5,775.90
0	W-1	4,332.60	4,543.80	4,543.80	4,751.70	4,751.70	4,915.50	4,915.50	5,065.80	5,065.80	5,248.80
Enlisted Members											
22	E-9	7,285.50	7,650.00	7,650.00	7,650.00	7,650.00	8,033.10	8,033.10	8,033.10	8,033.10	8,033.10
18	E-8	5,921.70	6,040.50	6,040.50	6,040.50	6,040.50	6,040.50	6,040.50	6040.50	6,040.50	6,040.50
13	E-7	4,940.40	4,940.40	5,291.40	5,291.40	5,291.40	5,291.40	5,291.40	5,291.40	5,291.40	5,291.40
6	E-6	3,944.10	3,944.10	3,944.10	3,944.10	3,944.10	3,944.10	3,944.10	3,944.10	3,944.10	3,944.10
3	E-5	3,310.50	3,310.50	3,310.50	3,310.50	3,310.50	3,310.50	3,310.50	3,310.50	3,310.50	3,310.50
2	E-4	2,596.50	2,596.50	2,596.50	2,596.50	2,596.50	2,596.50	2,596.50	2,596.50	2,596.50	2,596.50
1	E-3	2,176.80	2,176.80	2,176.80	2,176.80	2,176.80	2,176.80	2,176.80	2,176.80	2,176.80	2,176.80
0	E-2	1,836.30	1,836.30	1,836.30	1,836.30	1,836.30	1,836.30	1,836.30	1,836.30	1,836.30	1,836.30
0	E-1	1,638.30	1,638.30	1,638.30	1,638.30	1,638.30	1,638.30	1,638.30	1,638.30	1,638.30	1,638.30
0	E-1<4	0.00	0.00	0.00	0.00	0.00	0.00	0.00	0.00	0.00	0.00

SOURCE: Authors' calculations.

NOTE: Yellow highlighted cells are values that are not derived from the TIS monthly basic pay table (Table 2.1) but are imputed, as described in the main text.

Figure 2.1
Simulated Monthly Basic Pay over a Career, Time-in-Grade Versus Time-in-Service Pay Tables with Due-Course Promotion Histories, Enlisted Personnel (top left), Commissioned Officers (right), Warrant Officers (bottom left)

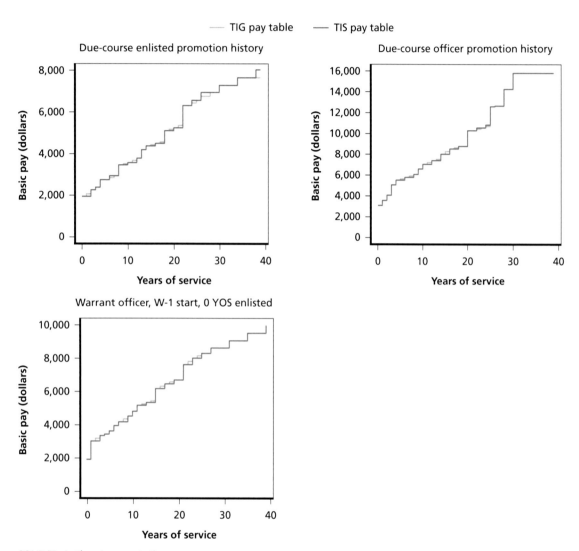

SOURCE: Authors' computations.
NOTE: The figure for warrant officers on the bottom left shows the pay profile for a member with no prior enlisted service who starts at grade W-1.

sonnel may be considered for promotion every year over a wide YOS interval. Because of this difference, we do not show pay over a career for slow promotion to O-4, just due-course versus one year faster. Pay over a career for an officer who is promoted faster than their year group is higher under the TIG pay table. To more clearly see the difference, we show in Figure 2.4 pay over a career for a fast-promoting officer under the TIG versus the TIS pay table. Pay is higher under the TIG pay table especially in the mid-career, but the difference in pay is not large. In large part, this relatively small difference reflects the structure of the officer pay table. As discussed in Asch (2019a), the officer pay table is relatively compressed in terms of differences in basic pay across grades. The construction of the TIG pay table for officers is built on the current officer pay table and so also reflects this compression. Thus, the main conclusion

Figure 2.2
Simulated Monthly Basic Pay over a Career, Time-in-Grade Versus Time-in-Service Pay Tables for Fast- Versus Slow-Promoting Enlisted Personnel

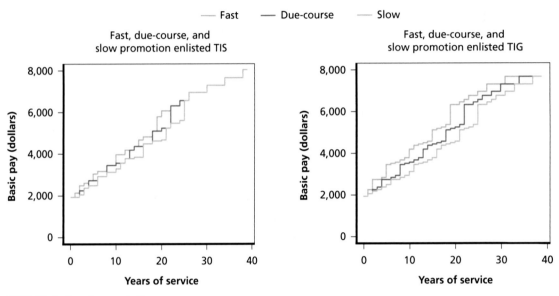

SOURCE: Authors' computations.

Figure 2.3
Simulated Monthly Basic Pay over a Career, Time-in-Grade Versus Time-in-Service Pay Tables for Fast- Versus Due-Course-Promoting Officers

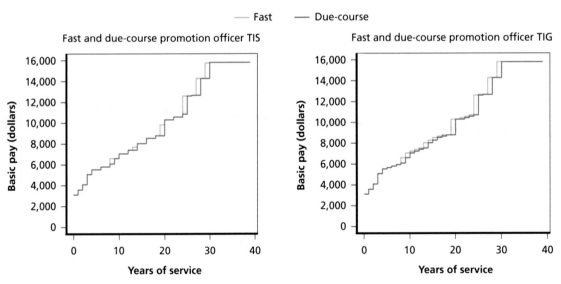

SOURCE: Authors' computations.

is that while pay is higher under the TIG table for an officer who is promoted faster, they pay advantage is not large. Figures 2.3 and 2.4 show results for fast promotion to O-4, and a similar result is found for fast promotion to O-5 (not shown).

Effects on Pay of Time in Grade for Members in Fast- Versus Slow-Promoting Occupations

A disadvantage of the TIG pay table discussed in past studies is that promotion timing differences, and therefore pay differences across services or across occupations within a service, may be a result of supply and demand conditions that are beyond the control of a given service member and not a result of differences in individual performance. While these differences occur under both a TIS and TIG pay table, the differences are magnified under the TIG pay table. The extent to which these differences vary across occupations within a service or across services is explored in Chapter Seven when we discuss the merits and drawbacks of the TIG pay system. Here, we illustrate the implications for pay over a career of differences in promotion timing across occupations within a given service.

Figure 2.5 shows basic pay over a career under the TIG versus the TIS pay table for slow- versus fast-promoting occupations. Occupations within DoD Occupation Code 7, Craftsmen, promote about one year slower than average to E-5 and E-6, whereas those within DoD Occupation 0, Infantry, Gun Crews, and Seamanship Specialists, promote about one year faster than average to E-5 and about two years faster than average to E-6 based on DMDC tabulations. The left panel of Figure 2.5 shows that basic pay is higher over a career under the TIG pay table versus the TIS pay table for those in DoD Occupation Code 0 (the faster-promoting occupation). Similarly, the right panel shows that basic pay over a career is lower under the TIG pay table for the slower-promoting occupation (DoD Occupation Code 7). The implication is that the TIG pay table provides a greater financial reward over a career for those in fast-promoting occupations and provides less of a reward for those in slow-promoting occupations than the TIS pay table.

Figure 2.4
Simulated Monthly Basic Pay over a Career, Time-in-Grade Versus Time-in-Service Pay Tables for Fast-Promoting Officers

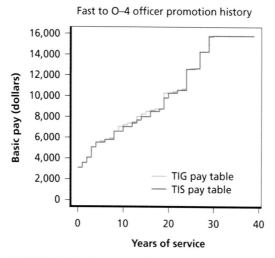

SOURCE: Authors' computations.

Figure 2.5
Simulated Monthly Basic Pay over a Career, Time-in-Grade Versus Time-in-Service Pay Tables for Fast- Versus Slow-Promoting Enlisted Occupations (DoD Occupation Codes 0 Versus 7)

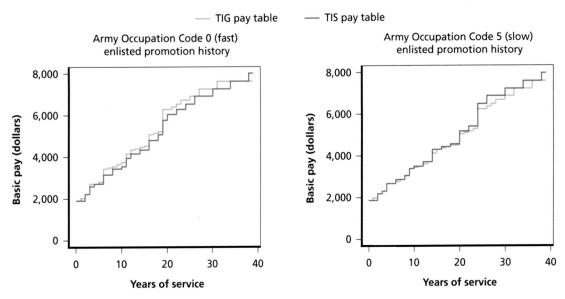

SOURCE: Authors' computations.

To assess the magnitude of the differences in Figure 2.5, we compute the discounted present value (DPV) of basic pay over a career in Table 2.3. Because the military retired pay formula is based on the highest-three YOS (typically YOS 18–20 for someone retiring at YOS 20), faster or slower promotion will affect the "high-3" computation of basic pay and thus the monthly retired pay annuity. Thus, differences in basic pay over a career can result in differences in retired pay, and so we show computations of the DPV of retired pay over a lifetime under the TIG versus the TIS pay table for fast- versus slow-promoting occupations shown in in Figure 2.4. The computation of retired pay assumes retirement at YOS 20. For the computations shown in Table 2.3, we use a retired pay multiplier of 2.5 percent, as under the legacy military retirement system.[3] Guided by estimates from past studies, we use a personal discount rate of 10 percent for the DPV computations.[4]

Faster-promoting occupations result in a higher DPV of basic pay under both the TIG and TIS pay table, but the difference is greater under the TIG table, 11.3 percent versus 5.5 percent. The differences in basic pay for fast- verses slow-promoting occupations translate into differences in the high-3 computation and, hence, the DPV of retired pay. As shown in the table, the differences are magnified under the TIG pay table, 22.8 percent versus 14.3 percent under the TIS pay table.[5]

---

[3]   Members entering service beginning 2018 are under the Blended Retirement System (BRS), in which the retired pay multiplier is 2.0 percent. Members with at most 12 YOS as of December 31, 2017, had the opportunity to opt into the BRS during calendar year 2018. The BRS includes three elements: a retirement annuity, a defined contribution plan, and continuation pay. Differences in basic pay associated with the TIG pay table could affect all three elements. The legacy military retirement was the system in effect prior to the introduction of the BRS.

[4]   These estimates are discussed in past RAND documents, such as in Asch, Hosek, and Mattock, 2014, Appendix E.

[5]   These relative differences also carry through under the BRS. While the absolute difference between the DPV of the monthly retirement pay annuity for fast- and slow-promoting occupations would be 20 percent smaller under the BRS

Table 2.3
Discounted Present Value of Enlisted Basic Pay and Retired Pay for Fast- and Slow-Promoting Occupations (2018 dollars)

Enlisted Personnel	TIS Pay Table	TIG Pay Table
Basic Pay		
Fast-promoting occupation (DoD Occupation Code 0)	$386,700	$404,400
Slow-promoting occupation (DoD Occupation Code 7)	$366,600	$363,300
Difference	5.5%	11.3%
Retired Pay		
Fast-promoting occupation (DoD Occupation Code 0)	$314,300	$334,300
Slow-promoting occupation (DoD Occupation Code 7)	$244,900	$272,200
Difference	14.3%	22.8%

SOURCE: Authors' computations.

NOTE: DoD Occupation 0 is Infantry, Gun Crews, and Seamanship Specialists, and DoD Occupation Code 7 is Craftsmen. Computations assume a 30-year military career and use a 10 percent personal discount rate. Retired pay computation is based on the legacy (pre-2018) military retirement formula equal to 2.5% of the highest-three years of basic pay times YOS and assumes an expected lifespan until age 85.

Warrant Officers with Prior Service

As noted by the 10th QRMC, warrant officers and commissioned officers with prior enlisted service present difficulties from a pay perspective in the current TIS pay table and greater difficulties in a TIG pay table. The difficulty is that members promoted from the enlisted force to either the warrant officer or commissioned officer force often have widely different amounts of prior enlisted service. This can result in a pay inversion or pay reduction at the time of transition to the officer corps for enlisted members. In this section, we focus on warrant officers, and we discuss commissioned officers in the next section.

Table 2.4 shows the grade and YOS eligibility requirements for warrant officers, by service. (For completeness, the table includes the Air Force, although the Air Force does not currently have a warrant officer program.) Except for Army aviators, warrant officers require prior enlisted service. The minimum YOS and grade requirements vary across service. For technical Army specialties (non-aviator), warrant officers must be at least an E-5 and have between 4 and 6 YOS. In contrast, Marine Corps warrant officers in nontechnical specialties require at least 16 YOS in the Marine Corps or 23 YOS in the Navy. Navy warrant officers must be at least an E-7 (or promotable as an E-6) with at least 12 YOS. As shown in the final column, the Navy allows more-senior enlisted personnel who become warrant officers to enter the warrant officer

---

(due to the BRS retired pay multiplier being 20 percent smaller), the relative difference would remain the same as that shown for the legacy retirement system, 22.8 percent under the TIG pay table versus 14.3 percent under a TIS pay table.

Table 2.4
Warrant Officer Grade and Years-of-Service Eligibility Requirements

Service	Career Fields	Minimum Enlisted Grade	Minimum YOS	Notes
Army	Technical	E-5	4–6 YOS	
	153A (aviators)	No prior service	Not applicable	
Marine Corps	Technical	E-5	8 YOS in USMC or 16 YOS in Navy	
	Nontechnical	E-7	16 YOS in USMC or 23 YOS in Navy	
Air Force	No Warrant Officer Program			
Navy	All	E-7 or E-6 (promotable)	12 YOS	Enter as W-2
Coast Guard	All	E-6	8 YOS with at least 4 in USCG	

SOURCES: Navy Personnel Command (undated), U.S. Marine Corps (2019), U.S. Army Recruiting Command (undated), U.S. Coast Guard (2017).

corps at the grade of W-2. This policy addresses the possibility of pay inversion whereby more senior enlisted personnel who become warrant officers may receive a pay cut.

Following the 10th QRMC, we developed the TIG pay table in Table 2.3 for warrant officers with entry grade points associated with a non-prior service warrant officer career, relevant only to Army aviators. As shown in the lower left panel of Figure 2.1, warrant officers with no prior service would have the same basic pay profile over a career under the TIG as under the TIS pay table, assuming these warrant officers received due-course promotions that are the same as the average promotion times between 2013 and 2018. However, a drawback of this TIG pay table design of warrant officers is that basic pay under the TIG pay table would be lower than under the TIS pay table for warrant officers with prior enlisted service. This is shown in Figure 2.6, in which basic pay over a career is shown under the TIG versus the TIS pay table for a member who transitions to warrant officers status after either 6 years or 12 years as an enlisted member (left and right panels, respectively). In the years prior to promotion to warrant officer, the figure shows basic pay during the enlisted portion of members' careers. By design, those receiving due-course (enlisted) promotions receive the same basic pay over the career under both the TIS and TIG tables, so the green and blue lines overlap in the figures. After promotion to warrant officer, pay is lower under the TIG pay table. In the case of those with 12 YOS as an enlisted member, pay is not only lower under the TIG pay table than under the TIS table, but for those under the TIG pay table, pay falls at the transition point to warrant officer status under the TIG pay table (but not the TIS table), as seen by the reduction in pay at 12 YOS relative to pay at 11 YOS in the right-hand panel.

Pay is lower under the TIG versus the TIS table for two reasons. First, the warrant officer TIG table was designed for those with no prior enlisted service. Second, unlike the TIS pay table, the TIG table does not account for YOS differences at the time of promotion to warrant officer. Put differently, the TIS pay table is distinctly more advantageous than the TIG table in ensuring that members who transition to warrant officer do not receive a pay cut.

One way to address the lower pay under the TIG pay table relative to the TIS table is to move the entry grade points in the TIG table to make them more senior. As discussed by the Hogan et al. (2008), the disadvantage of this approach is that pay for warrant officers without

Figure 2.6
Simulated Monthly Basic Pay over a Career, Time-in-Grade Versus Time-in-Service Pay Tables with Due-Course Promotion Histories for Warrant Officers with 6 or 12 Prior YOS as Enlisted

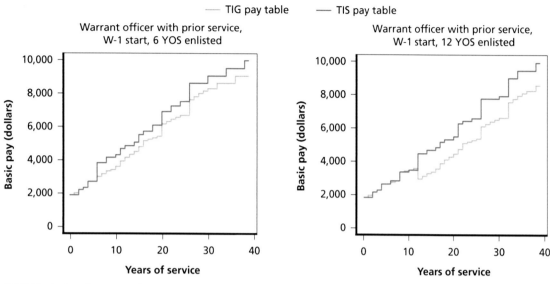

SOURCE: Authors' computations.

prior enlisted service would be substantially higher under the TIG relative to the TIS pay table. Another way is to allow warrant officers with substantial amounts of prior enlisted service to transition to warrant officer status at a grade higher than W-1. This is consistent with the Navy approach of allowing warrant officers to enter as a W-2.

Figure 2.7 shows basic pay over a career under the TIG pay table for warrant officers who transition at 12 YOS and are paid as a W-1 (left panel) versus as a W-2 (right panel) at the transition point. (To facilitate the comparison, the left panel in Figure 2.6 replicates the right panel from Figure 2.5.) While these warrant officers would still receive lower pay under the TIG than the TIS pay table, they would no longer receive a pay cut at the point of transition under the TIG table. That is, pay at YOS 12 would exceed pay at YOS 11 under the TIG pay table. Figure 2.8 shows that paying these members as a W-3 at promotion to warrant officer would go a long way toward closing the gap in basic pay under the TIG versus the TIS pay table. In short, reductions in pay for senior enlisted members who become warrant officers under the TIG pay table could addressed by allowing entry at grades above W-1.

Commissioned Officers with Prior Enlisted Service

As with warrant officers, members promoted from the enlisted force to the commissioned officer force can have differing amounts of prior enlisted service. The TIS pay table has the advantage that it accounts for YOS at promotion. Furthermore, officers commissioning with at least 4 years of prior enlisted service begin their officer career in the grades of O-1E, O-2E, and O-3E in the TIS pay table. Pay in these grades is higher than pay for O-1 to O-3, i.e., officers with no prior enlisted service. An important consideration in the design of the pay table for these officers is that they do not experience a pay cut as they transition from O-3E to O-4.

Figure 2.7
Simulated Monthly Basic Pay over a Career, Time-in-Grade Versus Time-in-Service Pay Tables with Due-Course Promotion Histories for Warrant Officers 12 Prior YOS as Enlisted, Entering as W-2

SOURCE: Authors' computations.

Figure 2.8
Simulated Monthly Basic Pay over a Career, Time-in-Grade Versus Time-in-Service Pay Tables with Due-Course Promotion Histories for Warrant Officers 12 Prior YOS as Enlisted, Entering as W-3

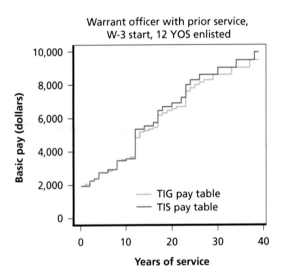

SOURCE: Authors' computations.

Because the TIS table accounts for YOS, it has the advantage that it is designed to provide a pay increment, and not a pay cut, to those promoted to O-4 from the grade of O-3E. But, the TIG table does not recognize the greater seniority of these commissioned officers and so they may experience a pay cut at the O-4 promotion point.

The design of the TIG pay table in Table 2.2 also includes grades O-1E, O-2E, and O-3E. To address the pay inversion issue that can arise at the O-4 promotion point, the design ensures that pay of an O-3E is always less than the pay of an O-4 while still increasing pay with more time in grade for those in the grade of O-3E. As shown by the highlighted cells for O-3E in Table 2.2, the pay for O-3E is imputed beyond 0 years in grade. The advantage of this design is that officers commissioned with prior enlisted service experience a pay increase at promotion from O-3E to O-4. A disadvantage is that these members experience slower pay growth as an O-3E than they would under a TIS pay table (a table that can directly recognize their greater experience).

Figure 2.9 shows basic pay growth over a career for members commissioned with 4 YOS as enlisted (left panel) or with 10 YOS at enlisted (right panel) under the TIG pay table design of Table 2.2. For officers commissioned with 4 YOS as an enlisted member (left panel), the first 4 YOS in Figure 2.9 reflect pay as an enlisted. At YOS 4, pay increases as the member transitions to the grade of O-1E. Because the pay of O-1E in the TIG table assumes entry at YOS 6, basic pay increases more under the TIG than the TIS pay table, i.e., the green line is above the blue line in the left panel of the figure. As an O-3E, beginning at YOS 8, pay increases are relatively flat. Pay increases are less flat once the member begins as an O-4 (at YOS 16 in the left panel), but pay growth is a bit slower until YOS 30 under the TIG than under the TIS table. We see a similar pattern for those commissioning after 10 YOS. These officers reach O-3E after 14 YOS and O-4 after 19 YOS. The structuring of O-3E pay to prevent pay inversion results in slower pay growth through YOS 40.

Figure 2.9
Simulated Monthly Basic Pay over a Career, Time-in-Grade Versus Time-in-Service Pay Tables for Commissioned Officers 4 Prior YOS (left) and 10 Prior YOS (right)

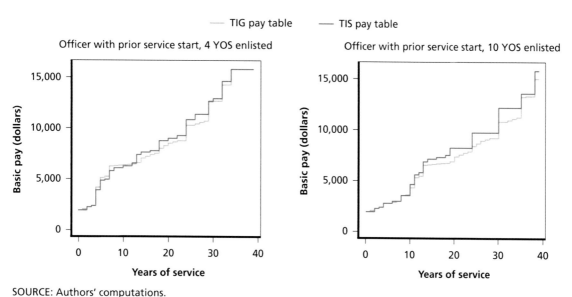

SOURCE: Authors' computations.

Lateral Entry

A major perceived advantage of a TIG pay table is that it can more easily facilitate the offering of higher pay to lateral entrants. As discussed in Chapter One, DOPMA reform included in the NDAA 2019 authorized the services to grant "constructive credit" for education as well as for work experience, thereby allowing individuals to enter service at a rank as high as an O-6 (colonel or Navy captain). To illustrate the advantage of the TIG pay table for facilitating lateral entry, we consider lateral entry as an O-4, consistent with the NDAA 2019 reforms. Figure 2.10 shows the simulation of basic pay over an officer career for individuals who enter military service as an O-4.

We find that pay is higher for lateral entrants under the TIG pay table. The reason is that lateral entrants receive the pay of an O-4 with 9 YOS, the entry YOS point for an O-4 (see Table 2.2) in the TIG table, rather than with 0 YOS, as would be the case under the TIS pay table.

That said, higher pay could also be offered under a TIS pay table if the definition of constructive credit were broadened to allow individuals to enter the military at both higher grades *and* YOS. Figure 2.11 shows the basic pay profile for O-4 lateral entrants under the TIS pay table, whereby individuals would receive constructive credit of 9 YOS. That is, at entry, these individuals would receive the pay of an O-4 with 9 YOS. The figure shows that the TIG pay table is no longer more advantageous in terms of providing higher pay to lateral entrants. In fact, pay is virtually identical under the TIG versus the TIS pay table. In Chapter Four, we return to the topic of policies that could be implemented under the current TIS pay table that might replicate the advantages of a TIG pay table.

Figure 2.10
Simulated Monthly Basic Pay over a Career, Time-in-Grade Versus Time-in-Service Pay Table for Lateral Entrant as an O-4

SOURCE: Authors' computations.

Figure 2.11
Simulated Monthly Basic Pay over a Career, Time-in-Grade with Constructive Credit Versus Time-in-Service Pay Table for Lateral Entrant as an O-4

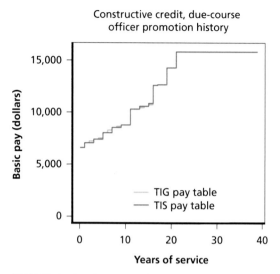

SOURCE: Authors' computations.

Summary

In this chapter, we developed a TIG pay table, extending the work of the 10th QRMC and the DACMC. The entry points or anchor points are critical inputs to the development of the pay table, and we used more recent promotion timing data than were used by the 10th QRMC. We simulated basic pay profiles over a career for enlisted personnel, warrant officers, and commissioned officers, and found that those who receive due-course promotions, or promotions that are exactly the same as the anchor points, receive the same pay over their career under the TIG as the TIS pay table. By providing a permanent pay increment or decrement to those promoted faster or slower, our simulations show that the TIG pay table shows more pay variability, and therefore greater incentives for performance, than the TIS pay table, insofar as promotion speed reflects performance.

A challenge with establishing the TIG pay table is the pay for warrant officers and commissioned officers who transition out of the enlisted force; the difficulty is that members promoted from the enlisted force often have widely different amounts of prior enlisted service. Another difficulty is that the warrant officer TIG pay table is designed for those without prior enlisted service. One way to address the lower pay under the TIG pay table relative to the TIS table is to move the entry grade points in the TIG table to make them more senior. The disadvantage of this approach is that pay for warrant officers without prior enlisted service would be substantially higher under the TIG relative to the TIS pay table.

Another way is to allow warrant officers with substantial amounts of prior enlisted service to transition to warrant officer status at a grade higher than W-1. A similar strategy could be used for commissioned officers. Finally, we found that the TIG pay table provides higher entry pay than the TIS pay table to lateral entrants. On the other hand, a similar result could be achieved under a TIS pay table if constructive credit were redefined so that entrants could receive pay at not only a higher grade but also a higher length of longevity.

CHAPTER THREE

Extending the Dynamic Retention Model to Analyze the Effect of a Time-in-Grade Pay Table

This chapter covers how we extended the DRM so that we can simulate the effect of a TIG pay table on retention, performance, and personnel costs. Performance is measured in terms of promotion speed relative to peers, where we consider two factors that can affect performance, ability, and effort supply. By *ability*, we mean characteristics of individual members that increase or decrease their promotion speed relative to their peers; these can include innate cognitive intelligence and other characteristics that lead to success, such as ability to work well in teams, ability to work in a hierarchical organizational structure, and resilience to changes such as frequent moves and new assignments. By *effort supply*, or simply *effort*, we refer to how hard and effectively members work in terms of achieving tasks that lead to faster promotion. In the simulations, we seek to provide estimates of the effect of the TIG pay table on overall retention, retention of individuals with higher innate ability, and the average ability and the level of effort exerted by individual members. As much of this chapter consists of technical material, readers whose main interest is in the policy analysis of the TIG pay table may wish to skip to the next chapter.

We first discuss how we extended the mathematical structure of the DRM to account for promotion. Explicitly modeling promotion is critical to being able to model a TIG pay table, because under this type of pay table compensation depends both on grade and the time since the last promotion. Previous versions of the DRM, with a few exceptions, modeled the military wage as being a function of YOS and did not explicitly model the promotion process. Given the expanded mathematical structure, we estimate the DRM parameters for enlisted personnel and officers in each service using DMDC data that track individual service members from entry in 1990 and 1991 through their active and reserve military career until 2016. We can then use the parameter estimates to simulate the effects of untried policies, such as the TIG pay table. Next, we discuss how we conduct these simulations. In particular, we discuss how we used the DRM mathematical structure, which is based on a TIS pay table and on historical career data for service members serving under a TIS pay table, to simulate the effect of implementing a TIG pay table. After that, we discuss how we extended the DRM to simulate how the different pay tables might affect the retention of members of differing levels of ability, where we assume that higher-ability members are promoted faster than their peers. Then we examine how the DRM can be extended to examine the effects of differing pay tables on the amount of effort an individual chooses to exert, when we assume that individuals who exert more effort will be promoted faster than their peers who exert less effort. We conclude the chapter with a short summary.

Extending the DRM Mathematical Structure to Account for Promotion

The DRM is a model of the service member's decision, made each year, to stay in or leave the active component and, for those who leave, to choose whether to participate in a reserve component and, if participating, whether to continue as a reservist. These decisions are structured as a dynamic program in which the individual seeks to choose the best career path, but the path is subject to uncertainty. The model is formulated in terms of parameters that are estimated with longitudinal data on retention in the active component and participation in the reserve component, and these data are then used to see how well the estimated model fits observed retention. We use the estimated parameters in policy simulations.

We have described the DRM in earlier documents in which we have estimated a DRM for officers and for enlisted personnel in each service and for selected communities, such as Air Force pilots and military mental health care providers (Asch et al., 2008; Mattock et al., 2016; Hosek et al., 2017). This chapter presents an overview of the DRM, describing the extension of the model to cover promotion for both enlisted and officers. The description presented in this chapter draws heavily on Asch et al. (2018).

In the DRM, a set of parameters underlies the individual member's retention decisions, and a goal of our analysis is to use individual-level data on active retention and reserve participation to estimate the parameters for both enlisted personnel and officers for all four services. We discuss the data we use in more detail later in this chapter, but, in short, we use the DMDC's Work Experience File (WEX) to track individual careers from 1990 to 2016.

Model Overview

In the behavioral model underlying the DRM, in each period the individual can choose to continue on active duty, leave the military to hold a job as a civilian, or leave the military to join a reserve component and hold a job as a civilian. The individual bases their decision on which alternative has the maximum value. The model assumes that an individual begins their military career in an active component.

Individuals are assumed to differ in their preferences for serving in the military. Each individual is assumed to have given, unobserved, preferences for active and reserve service, and these preferences do not change. The individual member, officer or enlisted, has knowledge of military pay and retirement benefits, as well as civilian compensation. In each period there are random shocks associated with each of the alternatives, and the shocks affect the value of the alternative. As shown next, the model explicitly accounts for individual preferences and military and civilian compensation, and, in this context, shocks represent current-period conditions that affect the value of being on active duty, being in the selected reserve while also being a civilian worker (or *reserve*, for short), or being a civilian worker and not in the reserve (*civilian* for short). Examples of what may contribute to a shock are a good assignment; a dangerous mission; an excellent leader; inadequate training or equipment for the tasks at hand; a strong or weak civilian job market; an opportunity for on-the-job training or promotion; the choice of location; a change in marital status, dependency status, or health status; the prospect of deployment or deployment itself; or a change in school tuition rates. These factors may affect the relative payoff of being in an active component, being in a reserve component, or being a civilian. The individual is assumed to know the distributions that generate the shocks, as well as the shock realizations in the current period but not in future periods.

Depending on the alternative chosen, the individual receives the pay associated with serving in an active component, working as a civilian, or serving in a reserve component while also working as a civilian. In addition, the individual receives the intrinsic monetary equivalent of the preference for serving in an active component or serving in a reserve component. These values are assumed to be relative to that of working as a civilian, which is set at 0.

In considering each alternative, the individual takes into account their current state and type. *State* is defined by whether the member is active, reserve, or civilian and by the individual's active YOS, reserve YOS, total years since first joining the military, pay grade, and random shocks.

Type refers to the level of the individual's preferences for active and reserve service. The individual recognizes that today's choice affects military and civilian compensation in future periods. Although the individual does not know when future military promotions will occur, they do know the promotion policy and can form an expectation of military pay in future periods. Further, the individual does not know what the realizations of the random shocks will be in future periods. The expected value of the shock in each state is 0. Depending on the values of the shocks in a future period, any of the alternatives—active, reserve, or civilian—might be the best at the time. Once a future period has been reached and the shocks are realized, the individual can reoptimize (i.e., choose the alternative with the maximum value at that time). The possibility of reoptimizing is a key feature of dynamic programming models that distinguishes them from other dynamic models. In the current period, with future realizations unknown, the best the individual can do is to estimate the expected value of the best choice in the next period, i.e., the expected value of the maximum. Logically, this will also be true in the next period, and the one after it, and so forth, so the model is forward-looking and rationally handles future uncertainty. Moreover, the model presumes that the individual can reoptimize in each future period, depending on the state and shocks in that period. Thus, today's decision takes into account the possibility of future career changes and assumes that future decisions will also be optimizing.

Mathematical Formulation

We denote the value of staying in the active component at time t as

$$V^S\left(k_t\right) = V^A(k_t) + \varepsilon_t^A,$$

where k_t is defined as

$$k_t = k_t\left(ay_t, ry_t, t, g_t\right),$$

or the vector of number of active years (ay_t) at time t, the number of reserve years (ry_t), total years since initial enlistment or accession, and grade (g_t). $V^A(k_t)$ is the nonstochastic value of the active alternative, and ε_t^A is a random shock.

The value of leaving at time t is

$$V^L\left(k_t\right) = \max\left[V^R\left(k_t\right) + \omega_t^R, V^C\left(k_t\right) + \omega_t^C\right] + \varepsilon_t^L,$$

where the member can choose between reserve (R) and civilian (C). *Civilian* means working at a nonmilitary job, and *reserve* means participating in a reserve component and working at a nonmilitary job. The value of reserve is given by $V^R(k_t) + \omega_t^R$ where k_t is defined above, while value of civilian is given by $V^C(k_t) + \omega_t^C$. We model the reserve/civilian choice as a nest and assume that the stochastic terms follow an extreme value type I distribution, which leads to a nested logit specification in the estimation phase of this structural model.[1] The within-nest shocks to the reserve/civilian choice are given by ω_t^R and ω_t^C, and the nest-level shock is given by ε_t^L.

We allow a common shock for the reserve and civilian nest, ε_t^L, since an individual in the reserves also holds a civilian job, as well as shock terms specific to the reserve and civilian states, ω_t^R and ω_t^C. The individual is assumed to know the distributions that generate the shocks and the shock realizations in the current period but not in future periods. The distributions are assumed to be constant over time, and the shocks are uncorrelated within and between periods. Once a future year is reached, and the shocks are realized, the individual can reoptimize, i.e., choose the alternative with the maximum value at that time. But in the current period, the future realizations are not known, so the individual assesses the future period by taking the expected value of the maximum, i.e., the expected value of civilian conditional on it being superior to that of reserve times the probability of that occurring, plus the expected value of reserve conditional on it being superior to civilian times the probability of that occurring. For instance, depending on the shocks and the compensation, there is some chance that $V^S(k_t)$ will be greater than $V^L(k_t)$, in which case $V^S(k_t)$ would be the maximum, and vice versa, and the individual makes an assessment of the expected value of the maximum, $\text{Emax}(V^S(k_t), V^L(k_t))$.

The extreme value distribution, denoted *EV*, has location parameter a and scale parameter b; the mean is $a + b\phi$, and the variance is $\pi^2 b^2/6$, where ϕ is Euler's gamma (~0.577). As we derived in past studies (Asch et al., 2008; Mattock et al., 2016), this implies

$$\varepsilon_t^{Leave} \sim EV\left[-\phi\sqrt{\lambda^2 + \tau^2}, \sqrt{\lambda^2 + \tau^2}\right]$$

$$\omega_t^R \sim EV\left[-\phi\lambda, \lambda\right]$$

$$\omega_t^C \sim EV\left[-\phi\lambda, \lambda\right]$$

$$\omega_t^L \sim EV\left[-\phi\tau, \tau\right],$$

where λ is the common scale parameter of the distributions of ω_t^R and ω_t^C, and τ is the scale parameter of the distribution of ε_t^L. In the nested structure of the model, leavers face a common shock for the "leave" nest, ε_t^L, as well as shocks for the reserve and civilian alternatives within the nest, ω_t^R and ω_t^C, which all together produce a leave shock distributed as extreme value type I, with location parameter $-\phi\sqrt{\lambda^2 + \tau^2}$ and scale parameter $\sqrt{\lambda^2 + \tau^2}$. The logit model requires that the scale parameters of the leave and stay shocks be equal, so we parameterize the model such that the stay scale parameter, which we denote κ, has the same value as the leave scale parameter, i.e., $\kappa = \sqrt{\lambda^2 + \tau^2}$.

---

[1]   See Train, 2009, for a discussion of the logit and nested logit specifications.

The values of the alternatives $V^A(k_t)$, $V^R(k_t)$, and $V^C(k_t)$ depend on the current pay for serving in an active component or working as a civilian, $W^A(k_t)$ or $W^C(k_t)$. The members' active pay is based on total years of active service, ay_t, as well as their grade, g_t.

Our model includes promotion. The model assumes that the timing and probability of promotion at each grade is the same across all officers and is the same across all enlisted. Variation in the timing and probability of promotion for an individual member is captured by the shock term. Promotion to a given grade occurs at a given number of YOS, but the probability of promotion differs by grade. Also, the probability of promotion is assumed to be invariant to policy change. Not being promoted decreases the value of continuing in the military and operates to decrease retention. Officers or enlisted members that are promoted can look ahead to future promotion gates, and their value of staying is higher than that of members that are not promoted.

The possibility of re-optimizing in future periods distinguishes dynamic programming models from other dynamic models. Re-optimization means that the individual can choose the best alternative in a period when its conditions have been realized, i.e., when the shocks are known. As mentioned, future realizations are unknown in the current period, and the best the individual can do is estimate the expected value of the best choice in the next period, i.e., the expected value of the maximum. This will also be true in the following period, and the one after it, and so forth, so the model is forward-looking and rationally handles future uncertainty. Thus, today's decision takes into account the possibility of future changes of state and assumes that future decisions will also be optimizing.

To be more specific, in developing a mathematical expression for the value of the value function $V^A(k_t)$, the DRM considers all possible future pathways, recognizing that each pathway depends on each probability of promotion to the next grade and year of service when promotion can occur. Thus, the DRM views an officer or enlisted member with a particular k_t as reasoning forward to identify the full set of possible future paths of staying or leaving. Then, the member reasons backward starting from the final stay/leave decision year, called year T.

For each possible k_T, the model assumes that the member considers whether to stay or leave. From the perspective of an earlier year t, the member's current year, there is no reason to commit to a decision at T, and in fact it would be short-sighted to do so, because the member would not be able to base the decision on information that will be revealed when T arrives, i.e., when the shocks in T are realized. Instead, the member at t develops a decision rule about whether to stay or leave at T, and that rule is to stay if the value of doing so is higher than the value of leaving, otherwise to leave. The service member can—in the context of the model—compute the expected value of making that optimal decision. Reasoning backward, this expression enters into the expression for the optimal stay/leave decision at $T-1$ and so on back year by year to t.

At t, the value of continuing in the military for a member at grade g (now shown as a superscript) is

$$V^S\left(k_t\right) = V^A(k_t) + \varepsilon_t^A = \gamma^A + W_t^{Ag} + \beta EMax\left(V^A(k_{t+1}) + \varepsilon_{t+1}^A, V^L(k_{t+1}) + \varepsilon_{t+1}^L\right) + \varepsilon_t^A,$$

where γ^A is the individual's taste for active duty, W_t^{Ag} is active duty pay, β is the personal discount factor, the ε terms are random shocks, and the operator *Emax* finds the expected value of the maximum of the terms $V^A\left(k_{t+1}\right)+\varepsilon_{t+1}^A$ and $V^L\left(k_{t+1}\right)+\varepsilon_{t+1}^L$. Each of these terms has a non-random term and a random term.

Consider shocks that have an extreme value distribution with a mode of 0 and a scale of kappa: $\varepsilon \sim EV[0,\kappa]$. With an extreme value shock, the quantity $a + \varepsilon$ is distributed as $EV[a,\kappa]$. The mean of this distribution equals the scale factor times Euler's gamma plus the mode: $\phi\kappa + a$, where $\phi \approx 0.577$. If the mode is transformed by subtracting $\phi\kappa$, then $a - \phi\kappa + \varepsilon$ is distributed as $EV[a - \phi\kappa,\kappa]$ with a mean of a. (This transformation is equivalent to assuming that the shocks are distributed as $EV[-\phi\kappa,\kappa]$, that is, that the shocks have mean 0 and scale kappa.) Also, if two quantities V^m and V^n have the form $a + \varepsilon$ and we subtract $\phi\kappa$ from each, their maximum has an extreme value distribution, namely,

$$Max\left(V^m,V^n\right) \sim EV\left[\kappa\ln\left(e^{V^m/\kappa} + e^{V^n/\kappa}\right) - \phi\kappa,\kappa\right].$$

The mean of this distribution is

$$\kappa\ln\left(e^{V^m/\kappa} + e^{V^n/\kappa}\right).$$

The mean is literally the expected value of the maximum. This result implies that

$$EMax\left(V^A\left(k_{t+1}\right)+\varepsilon_{t+1}^A, V^L\left(k_{t+1}\right)+\varepsilon_{t+1}^L\right) = k\ln\left(e^{V^A(k_{t+1})/\kappa} + e^{V^L(k_{t+1})/\kappa}\right).$$

To introduce promotion, we replace V^A with its expected value, where p is the probability of promotion:

$$V^A = p_{t+1}^{g+1} V^{A(g+1)} + \left(1 - p_{t+1}^{g+1}\right) V^{Ag}.$$

In those YOS where no promotion occurs (that is, in those YOS when promotion is not possible), the probability of promotion is zero. In years where promotion might occur (i.e., in those YOS when promotion is possible), the probability of promotion is assigned a value relevant for the grade. In general, not all eligible individuals get promoted, particularly in the senior grades; as a result, the probability of promotion is typically strictly less than 1.

For simplicity, we assume that civilian pay only depends on YOS (or years since initial active enlistment or accession, if the individual has left active service). If the member is a reservist, they earn the civilian wage plus reserve pay, $W^C(k_t) + W^R(k_t)$. As with active pay, reserve pay depends on total years, including prior active years as well as, of course, reserve years.

The tastes for active and reserve duty, γ^A and γ^R, represent the individual's perceived net advantage of holding an active or reserve position, relative to the civilian state. Other things equal, a higher taste for active or reserve service increases retention. The tastes are assumed to be constant over time but vary across individuals. Also, tastes for active and reserve service are not observed but are assumed to follow a bivariate normal distribution among active component entrants.

The nonstochastic (in the current period) values of the reserve choice and civilian choice can be written as

$$V^R\left(k_t\right) = \gamma_r + W^C\left(k_t\right) + W^R\left(k_t\right) + \beta E\left[\max\left[V^R\left(k_{t+1}\right) + \omega_r, V^C\left(k_{t+1}\right) + \omega_c\right]\right]$$

$$V^C\left(k_t\right) = W^C\left(k_t\right) + R\left(k_t\right) + \beta E\left[\max\left[V^R\left(k_{t+1}\right) + \omega_r, V^C\left(k_{t+1}\right) + \omega_c\right]\right],$$

where $R(k_t)$ in the civilian equation is the value of any active or reserve military retirement benefit for which the individual is eligible. The 2016 NDAA created a new military retirement system, known as the Blended Retirement System. Because our data cover retention decisions of personnel under the legacy retirement system, we use the formula for the legacy system for the purpose of our analysis given by

$$R\left(k_t\right) = 2.5\% \times ay_t \times W^A\left(k_t\right)$$

for the active retirement system where, in this formula, $W^A(k_t)$ is the highest three years of basic pay and is computed based on total active years, ay_t. For a member with 30 YOS, the multiplier $2.5\% \times ay_t$ is 75 percent, while it is 100 percent for a member with 40 YOS. (After 2007, the 75 percent cap on the multiplier was lifted, thereby permitting additional YOS beyond 30 to contribute to retired pay.)

The model has two switching costs, which enter the relevant value function as additive terms. *Switching cost* refers to a de facto cost reflecting the presence of constraints or barriers affecting the movement from particular states and periods to other states, relative to the movement that would otherwise have been expected from the expressions shown above for the values of staying and of leaving. Switching costs are not actually paid by the individual but, as estimated in the model, are a monetary representation of the constraints or barriers affecting the transition from one state to another at a given time. Further, a switching cost can be either negative or positive. A negative value implies a loss to the individual when changing from the current status to an alternative status, while a positive value implies a gain, or incentive, for the change. The first switching cost is a cost of leaving the active component before an officer or an enlisted member's active duty service obligation (ADSO) is completed, or an enlisted member's initial term of service is completed. This switching cost enters the value functions $V^R(k_t)$ and $V^C(k_t)$. The estimates, shown later, indicate that the switching cost has a negative value for all services, possibly reflecting the perceived cost of breaching the service contract. The second switching cost is a cost of switching into the reserve from the civilian state, and enters the value function $V^R(k_t)$. This cost could represent difficulty in finding a reserve position in a desired geographic location or an adverse impact on one's civilian job, e.g., from not being available to work on certain weekends or for two weeks in the summer or being subject to reserve call-up. Its estimated value is negative across all services.

Estimation Methodology

To estimate the DRM, we use the mathematical structure of the model together with assumptions on the distributions of tastes across members and the shock distributions. This allows us to derive expressions for the transition probabilities, given one's state, which are then used to compose an expression for the likelihood of each individual's years of active retention and reserve participation. Importantly, each transition probability is itself a function of the underlying parameters of the DRM. These are the parameters of the taste distribution, the shock distributions, the switching costs, and the discount factor. The estimation routine finds parameter values that maximize the likelihood.

The transition probability is the probability in a given period of choosing a particular alternative, i.e., active, reserve or civilian, given one's state. Because we assume that the model is first-order Markov,[2] that the shocks have extreme value distributions, and that the shocks are uncorrelated from year to year, we can derive closed-form expressions for each transition probability. For example, as Train (2009) shows, the probability of choosing to stay active at time t, given that the member is already in the active component, is given by the logistic form

$$\Pr\left(V^S > V^L\right) = \frac{e^{\frac{V^A}{\kappa}}}{e^{\frac{V^A}{\kappa}} + \left(e^{\frac{V^R}{\lambda}} + e^{\frac{V^C}{\lambda}}\right)^{\frac{\lambda}{\kappa}}}.$$

We omit the state vector k_t in each expression for clarity. We can also obtain expressions for the probability of leaving the active component and, having left, the probabilities of entering, or staying in, the reserve component in each subsequent year. To relate the DRM to one-period discrete choice models, we note that in a given period and for a given state and individual taste, the individual's value functions for staying and leaving have the same form as those of a random utility model (RUM). Similarly, for those who have left active duty, the choices of whether to enter the reserves or to remain in the reserves are also based on a RUM. More broadly, the reserve choice is nested in the choice to leave active duty, and the model has a nested logit form. (See Train [2009] for further discussion.) Of course, the DRM differs from a traditional RUM because the explanatory variables are value functions, not simple variables such as age and education, and the value functions are recursive.

The transition probabilities in different periods are independent and can be multiplied together to obtain the probability of any given individual's career profile of active, reserve, and civilian states that we observe in the data. Multiplying the career profile probabilities together gives an expression for the sample likelihood that we use to estimate the model parameters for using maximum likelihood methods.[3] Optimization is done using the Broyden-Fletcher-Goldfarb-Shanno (BFGS) algorithm, a standard hill-climbing method. We compute standard

---

[2] A first-order Markov assumption is that the probability of an event at time $t + 1$ only depends on the state at time t.

[3] This approach bears some resemblance to a (highly restricted) mixed logit model.

errors of the estimates using numerical differentiation of the likelihood function and taking the square root of the absolute value of the diagonal of the inverse of the Hessian matrix. To judge goodness of fit, we use parameter estimates to simulate retention profiles for synthetic individuals (characterized by tastes drawn from the taste distribution) who are subject to shocks (drawn from the shock distributions), then aggregate the individual profiles to obtain a force-level retention curve and compare it with the retention curve computed from actual data.

We estimate the following model parameters:

- the mean and standard deviation of tastes for active and reserve service relative to civilian opportunities, (e.g., μ_a, μ_r, σ_a, and σ_r)
- a common scale parameter of the distributions of ω_t^R and ω_t^C, λ, and a scale parameter of the distribution of ε_t^L, or t
- a switching cost incurred if the individual leaves active duty before completing the active duty service obligation or first term
- a switching cost incurred if the individual moves from "civilian" to "reserve."

In past DRM analyses, we also estimate a personal discount factor (see Asch, Hosek, and Mattock, 2014). We fixed the personal discount factor in this study because we found that the model fits were better and parameter estimates were more reasonable relative to our expectations based on past research.[4] We set the personal discount factor for officers equal to 0.94 and for enlisted personnel to 0.88, which are the values we have typically estimated for officers and enlisted in earlier work.

Once we have parameter estimates for a well-fitting model, we can use the logic of the model and the estimated parameters to simulate the active component cumulative probability of retention to each YOS in the steady state for a given policy environment, such as a change to the retired pay cap. By *steady state*, we mean when all members have spent their entire careers under the policy environment being considered. The simulation output includes a graph of the active component retention profile for officers and enlisted personnel by YOS. We can also produce graphs of reserve component participation and provide computations of costs, though we do not do so here. We show model fit by simulating the steady-state retention profile in the current policy environment and comparing it with the retention profile observed in the data.

Data

DMDC's WEX data contain person-specific longitudinal records of active and reserve service. WEX data begin with service members in the active or reserve component on or after September 30, 1990. Our analysis files include active component entrants in 1990 and 1991, who are followed through 2016, providing up to 26 years of data for the 1990 cohort and up to 25 years of data for the 1991 cohort. In constructing the officer samples, we exclude medical personnel and members of the legal and chaplain corps because their career patterns differ markedly from those of the rest of the officer corps, suggesting that analysis of retention for these personnel needs to be conducted separately. We also excluded officers with prior enlisted

---

[4] The personal discount factor equals $1/(1 + r)$ where r is the personal discount rate. For example, a personal discount factor of 0.88 corresponds to a discount rate r of 13.6 percent.

service. Because the WEX does not include U.S. Coast Guard personnel, our analysis excludes this service.

Another key source of data is information on civilian and military pay. For civilian pay opportunities for enlisted personnel, we used the 2007 median wage for full-time male workers with associate's degrees. For officers, we use the 2007 80th percentile of basic pay for full-time male workers with a master's degree in management occupations for civilian pay. The data are from the U.S. Census Bureau. Civilian work experience is defined as the sum of active years, reserve years, and civilian years since age 22, but here pay does not vary by other factors, such as years since leaving active duty. We used 2007 military pay tables. Military pay increases are typically across-the-board, with the structure of pay by grade and year of service remaining the same.[5] Therefore, we did not expect our results to be sensitive to the choice of year. Annual military pay for active members is represented by RMC for FY 2007, equal to the sum of basic pay, basic allowance for subsistence, basic allowance for housing, and the federal tax saved because the allowances are not taxed. Data on RMC and basic pay by grade and YOS are from the *Selected Military Compensation Tables*, also known as the Green Book (Office of the Under Secretary of Defense for Personnel and Readiness, Directorate of Compensation, 1980–2018). Reserve component members are paid differently from active component members, although the same pay tables are used. The method for computing reserve component annual pay is described in Asch, Mattock, and Hosek (2017). Military retirement benefits are related to the basic pay table, and we use the basic pay tables for 2007 for this computation.

We also required data on enlisted and officer promotion rates and promotion timing to each grade. Officer promotion rates were drawn from those used in Asch and Warner (1994), and promotion rates for enlisted and promotion timing data for both officers and enlisted were based on computations of average time in service at promotion by grade and service, for FY 1993 to 2008, from DMDC. We chose these years because sought promotion times that would be relevant to the 1990–1991 accession.

Model Estimates and Model Fits for Officers

Tables 3.1 and 3.2 show the estimated parameters and standard errors for the retention model of officers. To make the numerical optimization easier, we did not estimate most of the parameters directly but instead estimated the logarithm of the absolute value of each parameter, except for the taste correlation, for which we estimated the inverse hyperbolic tangent of the parameter. All of the parameters are statistically significant in the Navy and Air Force models, and all but the between-nest scale parameter are significant in the Army and Marine Corps models. To recover the parameter estimates, we transformed the estimates. Table 3.3 shows the transformed parameter estimates for each service. The estimates are denominated in thousands of 2007 dollars, except for the assumed discount rate and the taste correlation.

---

[5]   An exception was the structural adjustment to the basic pay table in FY 2000, which gave larger increases to mid-career personnel who had reached their pay grades relatively quickly (after fewer YOS). A second exception was the expansion of the basic allowance for housing, which increased in real value from FY 2000 to FY 2005. The costing analysis is in 2018 dollars.

Table 3.1
Parameter Estimates and Standard Errors: Army and Navy Officers

	Army		Navy	
	Estimate	Standard Error	Estimate	Standard Error
Log(Scale Parameter, Nest = τ)	−1.36	33.83	5.20	0.04
Log(Scale Parameter, Alternatives within Nest = λ)	4.69	0.03	3.40	0.06
Log(−1*Mean Active Taste = μ_a)	3.19	0.04	3.00	0.05
Log(−1*Mean Reserve Taste = μ_r)	5.63	0.05	4.01	0.05
Log(SD Active Taste = σ_a)	3.76	0.04	3.87	0.05
Log(SD Reserve Taste = σ_r)	5.26	0.05	3.88	0.06
Atanh(Taste Correlation = ρ)	0.67	0.02	0.94	0.01
Log(−1*Switch Cost: Leave Active <ADSO)	4.81	0.03	5.20	0.04
Log(−1*Switch Cost: Switch from Civilian to Reserve)	6.05	0.03	4.90	0.05
Personal Discount Factor β (Assumed)	0.94	N/A	0.94	N/A
−1*Log Likelihood	24,141		32,139	
N	5,318		6,445	

SOURCE: Parameter estimates from cohorts of personnel entering active duty as officers in 1990–1991.

NOTES: The scale parameter κ governs the shocks to the value functions for staying and for the reserve-versus-civilian nest and equals $\sqrt{\lambda^2 + \tau^2}$. The means and standard deviations of tastes for active and reserve service relative to civilian opportunities are estimated, as are the costs associated with leaving active duty before completing ADSO and switching from civilian status to participating in the reserves. The personal discount factor was assumed to be 0.94 in these models.

Table 3.2
Parameter Estimates and Standard Errors: Air Force and Marine Corps Officers

	Air Force		Marine Corps	
	Estimate	Standard Error	Estimate	Standard Error
Log(Scale Parameter, Nest = τ)	4.79	0.09	1.02	3.49
Log(Scale Parameter, Alternatives within Nest = λ)	3.96	0.35	4.37	0.05
Log(−1*Mean Active Taste = μ_a)	2.92	0.07	2.65	0.07
Log(−1*Mean Reserve Taste = μ_r)	6.20	0.53	4.93	0.08
Log(SD Active Taste = σ_a)	3.24	0.09	3.16	0.07
Log(SD Reserve Taste = σ_r)	5.78	0.55	4.51	0.08
Atanh(Taste Correlation = ρ)	0.45	0.01	0.56	0.04
Log(−1*Switch Cost: Leave Active <ADSO)	4.73	0.06	4.89	0.05
Log(−1*Switch Cost: Switch from Civilian to Reserve)	5.52	0.34	5.63	0.05
Personal Discount Factor β (Assumed)	0.94	N/A	0.94	N/A
−1*Log Likelihood	8,871		9,086	
N	2,339		1,757	

SOURCE: Parameter estimates from cohorts of personnel entering active duty as officers in 1990–1991.

NOTES: The scale parameter κ governs the shocks to the value functions for staying and for the reserve-versus-civilian nest and equals $\sqrt{\lambda^2 + \tau^2}$. The means and standard deviations of tastes for active and reserve service relative to civilian opportunities are estimated, as are the costs associated with leaving active duty before completing ADSO and switching from civilian status to participating in the reserves. The personal discount factor was assumed to be 0.94 in these models.

Table 3.3
Transformed Parameter Estimates: Officers

	Army	Navy	Air Force	Marine Corps
Scale Parameter, Nest = τ	0.26	181.83	120.73	2.78
Scale Parameter, Alternatives within Nest = λ	109.15	29.96	52.67	78.68
Mean Active Taste = μ_a	–24.30	–20.06	–18.51	–14.14
Mean Reserve Taste = μ_r	–279.98	–55.37	–490.71	–138.94
SD Active Taste = σ_a	42.89	47.77	25.50	23.53
SD Reserve Taste = σ_r	191.57	48.66	324.13	90.75
Taste Correlation = ρ	0.58	0.74	0.42	0.51
Switch Cost: Leave Active < ADSO	–122.34	–180.42	–113.49	–133.39
Switch Cost: Switch from Civilian to Reserve	–425.02	–133.41	–248.92	–277.81
Personal Discount Factor β (Assumed)	0.94	0.94	0.94	0.94

NOTE: Transformed parameters are denominated in thousands of 2007 dollars, with the exception of the taste correlation and personal discount factor. Definitions of variables are provided in the Table 3.1 notes.

The remaining paragraphs of this section are devoted to a service-by-service narrative exploring the meaning of the parameter estimates; readers more interested in how well the model fits the data may wish to skip to the next subsection, on model fit.

We found that mean active taste is negative for the Army and equal to –$24,300. A negative value is consistent with past studies estimating the mean active taste among military officers and suggests that the military must offer relatively high pay to compensate for the requirements of service on active duty relative to not being in the military. For the Navy, the point of estimate of mean active taste is negative but smaller in absolute value than for the Army, equal to –$20,060. The mean active taste is also smaller in absolute value for both the Air Force and Marine Corps, at –$18,510 and –$14,140, respectively. All estimates of mean active taste are statistically significantly different from zero.

Mean taste for reserve duty is negative: –$279,980 for Army officers, –$55,370 for Navy officers, –$490,710 for Air Force officers, and –$138,940 for Marine Corps officers. As for the variance in tastes, we found that the standard deviation of active-duty taste is larger for the Army and the Navy, at $42,890 for Army officers and $47,770 for Navy officers, while the standard deviation of active-duty taste is smaller for Air Force and Marine Corps officers, at $25,500 and $23,530 respectively. The standard deviation of reserve taste is largest for the Air Force at $324,130, followed by the Army at $191,570, the Marine Corps at $90,750, and the Navy at $48,660.

The estimated scale parameter for the between-nest shock in the Navy model is much larger than the means and standard deviations of tastes, while the within-nest shock is of the same order of magnitude. These scale parameters provide information on the standard deviation of the common random shock for the reserve/civilian nest, as well as the within civilian/reserve nest shocks. The model nests the reserve and civilian alternatives because most reservists also hold a civilian job; hence, a shock to civilian is also likely to be felt by reserve. The scale parameter for the active and reserve/civilian shock is $\sqrt{\lambda^2 + \tau^2}$, while the within civilian/reserve nest shock is λ. We estimate λ to be $29,960 and τ to be $181,830 for the Navy. These estimates imply that the scale parameter for the total shock, κ, is $184,278. The relative magnitudes of the scale parameters suggest that movement between the active nest and the reserve/

civilian nest is largely driven by random shocks rather than by diverse tastes among Navy members (i.e., taste heterogeneity), while the movement between civilian and reserve statuses are equally driven by diverse tastes and random shocks.

For the Air Force, we found that the between-nest shock τ is larger than the mean and standard deviation of active taste, but smaller in absolute value than the mean and standard deviation of reserve taste. We estimated a τ of \$120,730, about six times the absolute value of the active mean taste of –\$18,510 and about five times the standard deviation of the active taste of \$25,500. However, the estimated value of τ is about one-fourth of the absolute value of the reserve mean taste at –\$490,710 and about one-third of the standard deviation of reserve taste, \$324.13. The within-nest shock λ is estimated to be \$52,670, which, like the estimate for τ, places it between the absolute values of the estimates for the mean and standard deviation of active taste and the mean and standard deviation of reserve taste. The relative sizes of these parameters suggest that movement between the active nest and the reserve/civilian nest are driven by a combination of both members' individual tastes and random shocks.

For the Army, we found that τ is small and not statistically significantly different from zero, so that the scale parameter for the active and reserve/civilian shock is essentially reduced to λ. We estimated a λ of \$109,150, approximately four times the estimated mean active taste of –\$24,300, and about half the value of the (absolute value of the) estimated mean reserve taste of –\$191,570, implying that tastes, as well as shocks, play a role in explaining shifts into and out of active, reserve, and civilian statuses for the Army.

Similarly, for the Marine Corps we found that we found that τ is small and not statistically significantly different from zero. As a result, the scale parameter for the active and reserve/civilian shock is essentially reduced to λ. The estimated value of λ is \$78,680, significantly larger than the mean and standard deviation of active taste at –\$14,140 and \$23,530, respectively, and smaller than the mean and standard deviation of reserve taste at –\$138,940 and \$90,750, respectively.

The switching costs for leaving active-duty early, before completing ADSO, are –\$122,340 for Army officers, –\$180,420 for Navy officers, –\$113,490 for Air Force officers, and –\$133,390 for Marine Corps officers. The cost of switching to a reserve component after being a civilian is –\$425,020 for Army officers, –\$248,920 for Navy officers, –\$113,490 for Air Force officers, and –\$277,810 for Marine Corps officers. These high costs may reflect the difficulty of finding an available reserve position or an implicit cost to one's civilian career and lifestyle.

Model Fit for Officers

To assess model fit, we used the parameter estimates to simulate the behavior of 10,000 synthetic service members represented by tastes drawn from the active/reserve taste distribution and subject to shocks drawn from a shock distribution with a scale parameter equal to the estimated value. Given active and reserve tastes, current-period shock values, knowledge of the expected pay and promotion environment in the military and the civilian world, and knowledge of the shock scale parameter, each synthetic individual, behaving as a dynamic-program decisionmaker, makes a stay-or-leave decision in each YOS in the active component. This generates a career length of service in the active component. After leaving active service, the individual becomes a civilian and makes a yearly decision regarding reserve participation. If the individual is not in the reserves, the decision is whether to participate; if the individual is in the reserves, the decision is whether to continue to participate. These decisions generate information about reserve participation by year for the years after active component service. We obtained the predicted active component retention profile by adding together these simulated active component retention profiles across a large number of simulated individuals, and we similarly combined individual reserve participation profiles to obtain the predicted reserve participation profile for the population of simulated individuals. The predicted profiles are plotted against the actual profiles to assess goodness of fit.

Figures 3.1 through 3.4 show the model fit graphs for the active component for each of the four services. The red lines are simulated cumulative retention, and the black lines are retention observed in the data. The figures show the Kaplan-Meier survival curves, and the dotted lines show the 95 percent confidence intervals for the Kaplan-Meier estimates for the observed data. The horizontal axis counts years since the individual was observed beginning active service. The vertical axis shows the cumulative probability of retention on active duty until that year. For example, at entry, YOS is 0 and the fraction of personnel retained is 1, and the fraction of the force retained falls over an active career as officers leave active duty. The solid black line shows the actual retention of individuals in our cohorts, and the red line shows the predicted retention. The numbers beneath the x-axis correspond to the model parameters shown in Tables 3.1 or 3.2 and help to ensure that a given figure matches a particular set of estimates. We assess goodness of model fit by visual inspection, that is, in terms of how well the black and red lines coincide.

Visual inspection reveals that model fit for the active component is good for the Army, Air Force, and Marine Corps, and that the model captures the general sweep of Navy retention. In all cases, the simulated retention line lies close to the observed retention line and reflects the pattern of retention seen in the data with attrition first being high, then slowing after mid-career as vesting in the defined-benefit retirement approaches, and then falling quickly once the vesting point is reached.

Figure 3.1
Model Fit Results: Army Officers

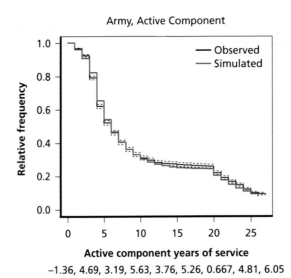

SOURCE: Authors' computations, DMDC WEX files.
NOTE: The numbers beneath the x-axis correspond to the model parameters shown in Table 3.1.

Figure 3.2
Model Fit Results: Navy Officers

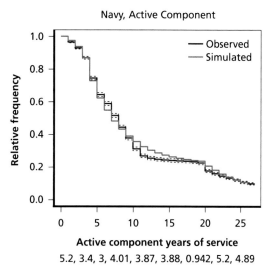

Navy, Active Component

5.2, 3.4, 3, 4.01, 3.87, 3.88, 0.942, 5.2, 4.89

SOURCE: Authors' computations, DMDC WEX files.
NOTE: The numbers beneath the x-axis correspond to
the model parameters shown in Table 3.1.

Figure 3.3
Model Fit Results: Air Force Officers

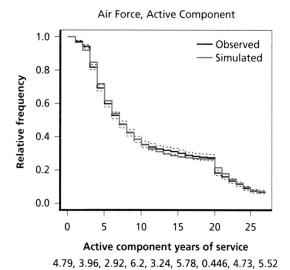

Air Force, Active Component

4.79, 3.96, 2.92, 6.2, 3.24, 5.78, 0.446, 4.73, 5.52

SOURCE: Authors' computations, DMDC WEX files.
NOTE: The numbers beneath the x-axis correspond to
the model parameters shown in Table 3.2.

Figure 3.4
Model Fit Results: Marine Corps Officers

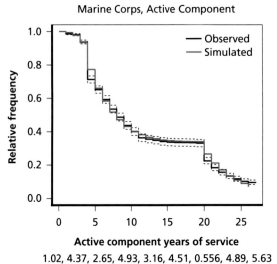

Marine Corps, Active Component

1.02, 4.37, 2.65, 4.93, 3.16, 4.51, 0.556, 4.89, 5.63

SOURCE: Authors' computations, DMDC WEX files.
NOTE: The numbers beneath the x-axis correspond to
the model parameters shown in Table 3.2.

Model Estimates and Model Fits for Enlisted Personnel

Tables 3.4 and 3.5 show the estimated parameters and standard errors for the enlisted DRM for the Army and Navy and Air Force and Marine Corps, respectively. As with the officer models, to make the numerical optimization easier, we did not estimate most of the parameters directly but instead estimated the logarithm of the absolute value of each parameter, except for the taste correlation, for which we estimated the inverse hyperbolic tangent of the parameter. All but the between-nest scale parameters τ are statistically significant in the models. To recover the parameter estimates, we transformed the estimates. Table 3.6 shows the transformed parameter estimates for each service. The estimates are denominated in thousands of 2007 dollars, except for the assumed discount rate and the taste correlation.

The remaining paragraphs of this subsection are devoted to a service-by-service narrative exploring the meaning of the parameter estimates; readers more interested in how well the model fits the data may wish to skip to next sub-section on model fit.

We found that mean active tastes are negative and equal to –$13,720, –$17,970, –$12,740, and –$44,650 for the Army, Navy, Air Force, and Marine Corps, respectively. The negative values are consistent with past studies and suggest that the military must pay a relatively high wage to compensate for the rigors of military life and retain enlisted members. All estimates of mean active taste are statistically different from zero.

The mean reserve tastes are also negative and are equal to –$24,100, –$26,580, –$165,070, and –$1,665,980 for the Army, Navy, Air Force, and Marine Corps, respectively. As for the variance in tastes, we found that the standard deviation of active-duty taste is largest for the Marine Corps at $28,100, while the standard deviation of active-duty taste is smaller for Army, Navy, and Air Force enlisted members, at $3,010, $6,880, and $7,590 respectively. Similarly, the standard deviation of reserve taste is largest for the Marine Corps at $1,113,030, followed by the Air Force at $109,510, the Army at $13,450, and the Navy at $13,150.

Table 3.4
Parameter Estimates and Standard Errors: Army and Navy Enlisted

	Army		Navy	
	Estimate	Standard Error	Estimate	Standard Error
Log(Scale Parameter, Nest = τ)	2.91	0.06	2.94	0.05
Log(Scale Parameter, Alternatives within Nest = λ)	2.39	0.11	1.70	0.10
Log(–1*Mean Active Taste = μ_a)	2.62	0.02	2.89	0.04
Log(–1*Mean Reserve Taste = μ_r)	3.18	0.10	3.28	0.10
Log(SD Active Taste = σ_a)	1.10	0.20	1.93	0.11
Log(SD Reserve Taste = σ_r)	2.60	0.12	2.58	0.12
Atanh(Taste Correlation = ρ)	0.68	0.03	0.26	0.02
Log(–1*Switch Cost: Leave Active <ADSO)	2.68	0.06	2.82	0.07
Log(–1*Switch Cost: Switch from Civilian to Reserve)	3.87	0.11	3.13	0.10
Personal Discount Factor β (Assumed)	0.88	N/A	0.88	N/A
–1*Log Likelihood	24,712		16,184	
N	5,540		4,863	

SOURCE: Parameter estimates from cohorts of enlisted personnel entering active duty in 1990–1991.

NOTES: The scale parameter κ governs the shocks to the value functions for staying and for the reserve versus-civilian nest and equals $\sqrt{\lambda^2 + \tau^2}$. The means and standard deviations of tastes for active and reserve service relative to civilian opportunities are estimated, as are the costs associated with leaving active duty before completing ADSO and switching from civilian status to participating in the reserves. The personal discount factor was assumed to be 0.88 in these models. Army and Navy models were estimated using a 5% random sample of the data.

Table 3.5
Parameter Estimates and Standard Errors: Air Force and Marine Corps Enlisted

	Air Force		Marine Corps	
	Estimate	Standard Error	Estimate	Standard Error
Log(Scale Parameter, Nest = τ)	0.23	4.04	0.41	8.45
Log(Scale Parameter, Alternatives within Nest = λ)	3.19	0.05	2.98	0.07
Log(–1*Mean Active Taste = μ_a)	2.54	0.03	3.80	0.04
Log(–1*Mean Reserve Taste = μ_r)	5.11	0.15	7.42	0.24
Log(SD Active Taste = σ_a)	2.03	0.10	3.34	0.06
Log(SD Reserve Taste = σ_r)	4.70	0.15	7.01	0.24
Atanh(Taste Correlation = ρ)	0.49	0.01	0.43	0.00
Log(–1*Switch Cost: Leave Active <ADSO)	2.98	0.06	4.13	0.05
Log(–1*Switch Cost: Switch from Civilian to Reserve)	4.80	0.05	4.28	0.08
Personal Discount Factor β (Assumed)	0.88	N/A	0.88	N/A
–1*Log Likelihood	10,313		11,251	
N	2,576		4,442	

SOURCE: Parameter estimates from cohorts of enlisted personnel entering active duty in 1990–1991.

NOTES: The scale parameter κ governs the shocks to the value functions for staying and for the reserve-versus-civilian nest and equals $\sqrt{\lambda^2 + \tau^2}$. The means and standard deviations of tastes for active and reserve service relative to civilian opportunities are estimated, as are the costs associated with leaving active duty before completing ADSO, and switching from civilian status to participating in the reserves. The personal discount factor was assumed to be 0.88 in these models. Air Force and Marine Corps models were estimated using a 5% and 10% random sample of the data, respectively.

Table 3.6
Transformed Parameter Estimates: Enlisted

	Army	Navy	Air Force	Marine Corps
Scale Parameter, Nest = τ	18.36	18.87	1.26	1.51
Scale Parameter, Alternatives within Nest = λ	10.95	5.45	24.38	19.70
Mean Active Taste = μ_a	−13.72	−17.97	−12.74	−44.65
Mean Reserve Taste = μ_r	−24.10	−26.58	−165.07	−1,665.98
SD Active Taste = σ_a	3.01	6.88	7.59	28.10
SD Reserve Taste = σ_b	13.45	13.15	109.51	1,113.03
Taste Correlation = ρ	0.59	0.25	0.46	0.40
Switch Cost: Leave Active < ADSO	−14.61	−16.73	−19.77	−62.16
Switch Cost: Switch from Civilian to Reserve	−48.12	−22.82	−122.11	−72.31
Personal Discount Factor β (Assumed)	0.88	0.88	0.88	0.88

NOTE: Transformed parameters are denominated in thousands of 2007 dollars, with the exception of the taste correlation and personal discount factor. Definitions of variables are provided in the Table 3.4 notes.

The estimated scale parameters for the between-nest shock in the Army and Navy are $18,360 and $18,870 respectively and are similar in size to the absolute value of the mean active and reserve taste parameters, while within-nest shock parameters for the Army and Navy at $10,950 and $5,450 are smaller than the absolute value of the mean taste parameters. The size of these scale parameters suggest that movement between the active nest and the reserve/civilian nest tends to be driven both by shocks and differences in tastes among enlisted members, while movement between civilian and reserve status tends to be driven more by taste. In the models for the Air Force and Marine Corps, the estimated scale parameter for the between-nest shock is much smaller than the means and standard deviations of tastes, at $1,260 and $1,510, respectively, and in both cases is not significantly different from zero, while the within-nest shock, at $24,380, and $19,700, is of the same order of magnitude as the absolute values of the active taste parameters, and uniformly smaller than the absolute values of the reserve taste parameters. The relative magnitudes of the scale parameters suggest that movement between the active nest and the reserve/civilian nest is equally driven by random shocks and diverse tastes among enlisted members, while the movement between civilian and reserve statuses tend to be more driven by taste than by random shocks.

The switching costs for leaving active-duty early, before completing the first term, are −$14,610 for Army enlisted members, −$16,730 for Navy enlisted members, −$19,770 for Air Force enlisted members, and −$62,160 for Marine Corps enlisted members. The cost of switching to a reserve component after being a civilian is −$48,120 for Army enlisted members, −$22,820 for Navy enlisted members, −$122,110 for Air Force enlisted members, and −$72,310 for Marine Corps enlisted members. These high costs may reflect the difficulty of finding an available reserve position within traveling distance of where the former active member has settled down.

Model Fit for Enlisted

Similar to the models of officer retention behavior, to assess model fit, we used the parameter estimates to simulate the behavior of synthetic personnel represented by tastes drawn from the active/reserve taste distribution and subject to shocks drawn from a shock distribution with a scale parameter equal to the estimated value. Figures 3.5 through 3.8 show the model fit graphs for the active component for each of the four services. The red lines are simulated cumulative retention, and the black lines are retention observed in the data. The figures show the Kaplan-Meier survival curves, and the dotted lines show the 95 percent confidence intervals for the Kaplan-Meier estimates for the observed data.

The horizontal axis counts years since the individual was observed beginning active service. The vertical axis shows the cumulative probability of retention on active duty until that

Figure 3.5
Model Fit Results: Army Enlisted

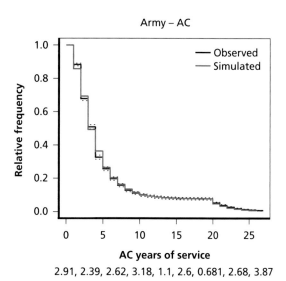

2.91, 2.39, 2.62, 3.18, 1.1, 2.6, 0.681, 2.68, 3.87

SOURCE: Authors' computations, DMDC WEX files.

Figure 3.6
Model Fit Results: Navy Enlisted

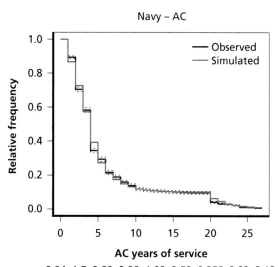

2.94, 1.7, 2.89, 3.28, 1.93, 2.58, 0.258, 2.82, 3.13

SOURCE: Authors' computations, DMDC WEX files.

Figure 3.7
Model Fit Results: Air Force Enlisted

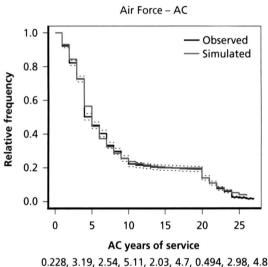

0.228, 3.19, 2.54, 5.11, 2.03, 4.7, 0.494, 2.98, 4.8

SOURCE: Authors' computations, DMDC WEX files.

Figure 3.8
Model Fit Results: Marine Corps Enlisted

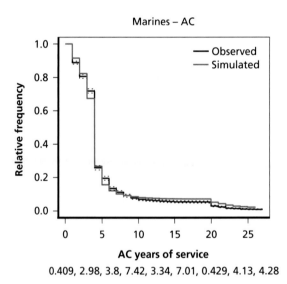

0.409, 2.98, 3.8, 7.42, 3.34, 7.01, 0.429, 4.13, 4.28

SOURCE: Authors' computations, DMDC WEX files.

year. The solid black line shows the actual retention of individuals in our cohorts, and the red line shows the predicted retention.

Visual inspection shows that the model fit for the active component is good for the Army, Navy, and Air Force, and that the model slightly over-predicts retention for the Marine Corps beyond YOS 10. In all cases the simulated retention line lies close to the observed retention line and reflects the pattern of retention seen in the data with attrition first being high, then slowing after mid-career as vesting in the defined-benefit retirement approaches, and then falling quickly once the vesting point is reached.

Simulation and Extension of the DRM to Model a Time-in-Grade Pay Table

To simulate the effect on retention of changing to a TIG pay table, we need to extend the DRM in two ways: (1) adapt the model to track time in grade, i.e., the number of YOS since a member was last promoted, and (2) ensure that military pay in the model is based on TIG rather than TIS.

The DRM was estimated using data on the behavior of officer and enlisted members under a TIS pay table, where the compensation an individual received was a function of their grade and YOS, which could conceptually be written as

$$W_t^{Ag} = W\left(ay_t, g_t\right).$$

Under a TIG pay table, the compensation a member receives is a function of their grade and the number of YOS since they were promoted to that grade. If we let py_t be the number of YOS since a member was last promoted, then we can write their wage as

$$W_t^{Ag} = W\left(py_t, g_t\right).$$

If we change the definition of k_t by adding py_t as follows

$$k_t = k_t\left(ay_t, ry_t, t, g_t, py_t\right),$$

then the rest of the mathematical expressions we developed earlier in this chapter still follow through. As a result, we can use the parameters estimated with the historical career data and TIS pay table to simulate the retention effects of replacing the TIS pay table with the TIG pay table. We also simulate the effects on performance and cost. We discuss how we incorporate performance in the next section. With respect to cost, we compute the total personnel cost per member of the simulated force produced under the TIS versus TIG pay table. Our estimates of personnel costs include the cost of basic pay, allowances, and the retirement accrual costs associated with the legacy military retirement system.

Incorporating Performance into the Dynamic Retention Model Simulation Capability

A major impetus for considering a TIG pay table is that it increases the incentives for performance, as discussed in Chapter Two. We incorporate performance into analysis by focusing on two aspects of individual service members that can affect their performance in the military: innate ability and how hard they work. This focus on the inputs of performance on the part of the member is consistent with two of the key objectives of the military compensation system related to individual performance: (1) to motivate personnel to work hard and effectively and

(2) to induce higher-ability personnel to stay and seek advancement to more-senior grades where it is likely that ability has a bigger impact than in the lower ranks.[6]

Asch and Warner were the first to incorporate ability and effort supply into a dynamic retention model, and they used the model to assess the retention, performance, and cost effects of alternative retirement reform proposals, as well as policies to restructure the military pay table (Asch and Warner, 1994a, 1994b, 2001). In particular, in their model, higher-ability personnel and those who exert more effort are promoted faster and have higher promotion probabilities, but higher-ability personnel also have better external opportunities, and expending effort involves a cost or disutility to the member (under the assumption that individuals would prefer to exert less effort for the same amount of financial benefit or return to effort). For higher-ability personnel, compensation policy can affect the financial returns to exerting more effort and the financial benefits to staying. Asch and Warner used their DRM to provide simulations of how compensation reforms affected overall retention, the retention of higher-ability personnel, ability sorting into higher grades, average effort supply, and personnel cost.

The Asch and Warner simulations were based on a calibrated model in which key parameters, such as the mean and standard deviation of taste for service, were assumed so as to replicate the observed retention profile. In contrast, the parameters of the DRM used in this study are estimated, not calibrated. We build on the Asch and Warner modeling of ability and effort and incorporate their approach into our DRM simulation capability to evaluate the TIS versus a TIG pay table. Ideally, we would consider both effort and ability simultaneously as factors affecting promotion probabilities, an approach taken by Asch and Warner. But we found that we were better able to incorporate ability and effort by considering them separately, as we'll discuss in more detail below. In the rest of this section, we first discuss how we incorporate ability and then effort.

Ability

We can use the structure of the DRM along with the estimated parameters and assumptions about how innate ability affects the speed of promotion to examine how selective the TIG and TIS pay tables are on ability. To incorporate ability into the DRM, we make assumptions about the following:

1. the extent to which ability differs among military entrants[7]
2. the extent to which ability affects promotion speed[8]
3. the effect of ability on external civilian opportunities.

We discuss each of these in turn.

---

[6]   The objectives of the military compensation are listed in DoD (2018) and have been articulated by past QRMCs and the DACMC.

[7]   We assume that the distribution of ability at entry is fixed and the same under a TIS and TIG pay table. Because we do not consider the effects of a TIG pay table on recruiting in this study, we do not consider the possibility that a TIG pay table might be more attractive to higher-ability recruits, thereby shifting the mean of the ability distribution. The implication is that a TIG pay table could have a greater effect on ability of the force than what we consider in this analysis.

[8]   The model only considers individual attributes in promotion timing/probability, so it does not allow for the possibility of the ability distribution skewing higher under TIG resulting in slowing down the promotion of individuals who might have been promoted early under TIS.

First, we assume that any given individual has a fixed level of ability at entry, drawn from a normal distribution and rounded to the nearest integer. The standard deviation of the distribution indicates the extent to which ability differs among military entrants. Regarding rounding, individuals with ability drawn from a normal distribution with mean 0 and standard deviation 0.5 (and then rounded) would typically have values of ability of –1, 0, or 1. We assume a different mean and standard deviation for each service and for enlisted personnel and for officers within that service. The values of the mean and standard deviation for each distribution we use in our simulations are calibrated to replicate the steady-state retention profiles of enlisted personnel and officers under the baseline TIS pay table, given the other two assumptions we make.

Second, we assume that higher-ability personnel are promoted faster. We implement this concept by subtracting the (rounded) draw from the normal distribution for a given individual from the TIS between promotions. This increase in promotion speed is modeled to start happening between E-5 and E-6 for enlisted members and between O-3 and O-4 for officers. Thus, an enlisted member with an innate ability of 1 would be one year faster than average to E-6, two years faster to E-7, and so on. An officer with an innate ability of 1 would be one year faster to O-4, two years faster to O-5, and so on. Consequently, the effect of ability on promotion speed to the more senior grades is larger than for the more junior grades because the effects on promotion timing are cumulative. Figure 3.9 shows how years to promotion to E-6 to E-9 vary with ability for Army enlisted personnel, and Figure 3.10 shows how years to promotion to O-4 to O-7 vary with ability for Army officers. Results will differ for the other services insofar as the assumed parameters of the ability distribution differ. As mentioned in the previous paragraph, the assumed parameters are calibrated so as to best fit the retention profile for that service and grade category.

Third, we assume that higher-ability members also have better external opportunities. We model this by multiplying the civilian opportunity wage by 1 plus 0.1 times the ability distribution standard deviation times the individual's ability draw, or $(1 + 1 \times \sigma_a)$ where σ_a is the standard deviation of the draw. This has the effect of increasing the civilian opportunity

Figure 3.9
Years to Promotion by Ability Level, Army Enlisted Personnel

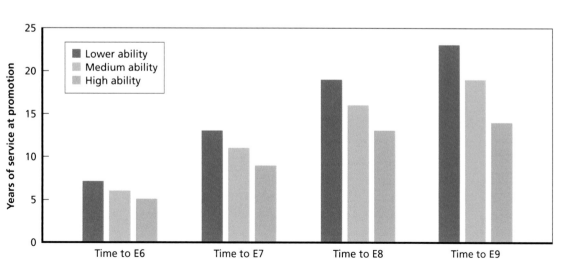

SOURCE: Authors' computations.

Figure 3.10
Years to Promotion by Ability Level, Army Officers

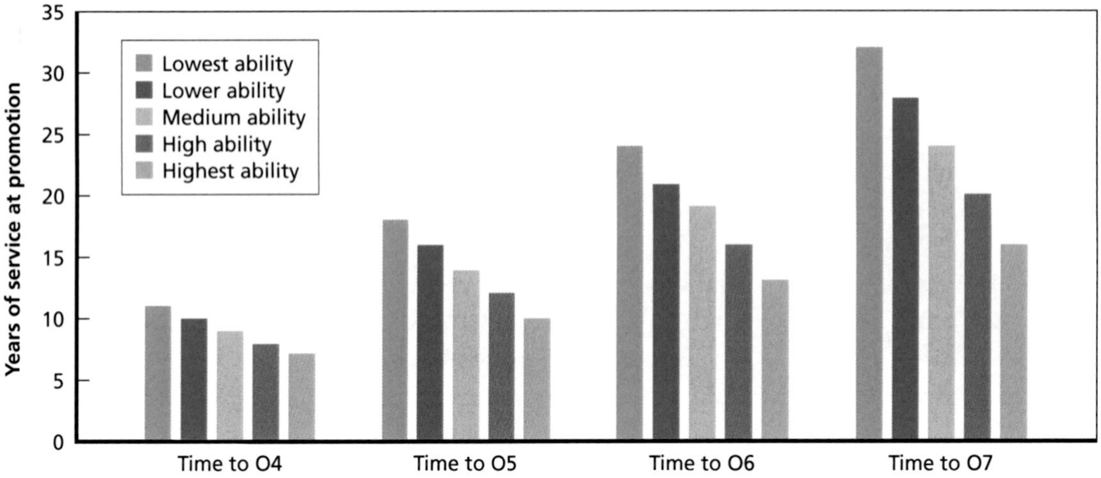

SOURCE: Authors' computations.

wage for high-ability individuals and decreasing the civilian opportunity wage for low-ability individuals. For example, an individual with innate ability of 1 drawn from a normal distribution with mean 0 and standard deviation 0.5 would have an opportunity wage that is 5 percent greater than that of the average individual, while an individual with innate ability –1 would face a civilian opportunity wage that is 5 percent less.

We illustrate how we calibrate the mean and standard deviation of the normal distribution to fit the observed retention profile in Figure 3.11 for Army enlisted personnel. In the process of calibration, we systematically varied the mean and standard deviation within the TIS DRM and chose the mean and standard deviation that most closely replicated the historically observed retention, as indicated by the Kaplan-Meier curve. The right panel shows the observed retention profile versus the simulated retention profile when we mis-calibrate the mean and standard deviation to equal 0 and 1.5, respectively. The simulated retention profile is too high relative to the observed profile. We chose a standard deviation of 0.5 instead resulting in a good fit, as shown in the left panel.

The three assumptions we make regarding how ability enters the model could affect our simulation results and in particular the effects of the TIG pay table on retention, ability sorting and cost. Consequently, our presentation of the results in Chapter Four includes sensitivity analyses in which we vary these three underlying assumptions regarding ability.

Modeling Effort

In addition to native ability, a member's promotion performance can depend on the amount of effort they exert. The main idea is that, other things held constant, the more effort a member exerts, the more likely they will be promoted. The structure of the model allows us to derive the optimal amount of effort an individual would exert given assumptions about how effort affects the probability of an individual being promoted, and assumptions about the disutility of effort.

Figure 3.11
Calibrating the Parameters of the Ability Distribution, Army Enlisted Personnel

Calibrated Mean = 0, Std. dev. = 0.5	**Mis-calibrated** Mean = 0, Std. dev. = 1.5

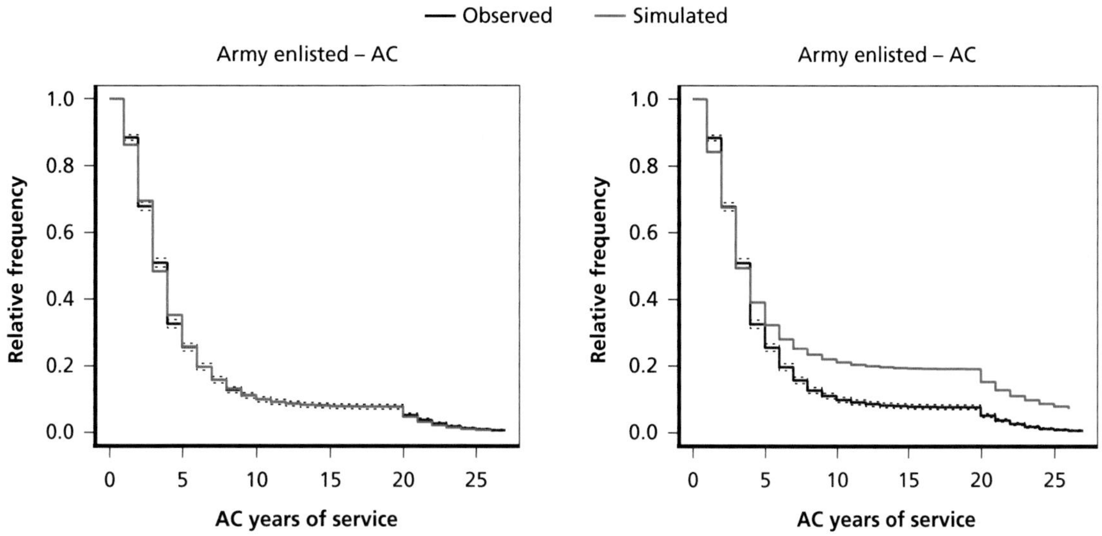

SOURCE: Authors' computations.

Following Asch and Warner (1994), we add disutility of effort to the value function in the DRM presented above. The individual's problem is to choose the level of effort to exert in the current period to maximize their utility:

$$\max_{e_t} V^A(k_t) - Z(e_t).$$

To simplify notation, we define $\bar{V}^A(k_t)$ to be the value of staying in the active component net the disutility of effort, like so:

$$\bar{V}^A(k_t) \equiv V^A(k_t) - Z(e_t).$$

The first-order condition for the optimal level of effort is

$$\frac{\partial \bar{V}^A(k_t)}{\partial e_t} = \beta \Pr\left(\bar{V}^S(k_{t+1}) > V^L(k_{t+1})\right)\left(\bar{V}^{A(g+1)}(k_{t+1}) - \bar{V}^{Ag}(k_{t+1})\right)\frac{\partial p_{t+1}^{g+1}}{\partial e_t} - Z'(e_t) \equiv 0.$$

or

$$\Pr\left(\bar{V}^S(k_{t+1}) > V^L(k_{t+1})\right)\beta\left(\bar{V}^{A(g+1)}(k_{t+1}) - \bar{V}^{Ag}(k_{t+1})\right)\frac{\partial p_{t+1}^{g+1}}{\partial e_t} \equiv Z'(e_t).$$

The interpretation of this expression is that the product of the probability of staying in the next period, the discounted difference of the value of being active and promoted and the value being active and not promoted, and the marginal effect of effort on the probability of promotion equals the marginal disutility of effort. Or, to put it more simply, the expected marginal return to effort equals the marginal disutility of effort.

If we make some assumptions regarding the functional form of the disutility of effort function and the probability of promotion as a function of effort, we can solve for optimal effort at time t. Similar to Asch and Warner, we let the disutility of effort be

$$Z(e_t) = \frac{\eta_0}{2} e_t^2$$

and let the probability of promotion be

$$p_{t+1}^{g+1} = \mu^{g+1} \overline{p}_{t+1}^{g+1} e_t,$$

where μ^{g+1} is a parameter that captures the relationship between effort and the probability of promotion for a given individual and $\overline{p}_{t+1}^{g+1}$ is the average promotion probability to grade $g + 1$ at time $t + 1$. We can rewrite the first-order condition as[9]

$$\beta \Pr\left(\overline{V}^S\left(k_{t+1}\right) > V^L\left(k_{t+1}\right)\right)\left(\overline{V}^{A(g+1)}\left(k_{t+1}\right) - \overline{V}^{Ag}\left(k_{t+1}\right)\right)\mu^{g+1}\overline{p}_{t+1}^{g+1} - \eta_0 e_t \equiv 0$$

and solve for e_t as:

$$e_t = \frac{\beta \Pr\left(\overline{V}^S\left(k_{t+1}\right) > V^L\left(k_{t+1}\right)\right)\left(\overline{V}^{A(g+1)}\left(k_{t+1}\right) - \overline{V}^{Ag}\left(k_{t+1}\right)\right)\mu^{g+1}\overline{p}_{t+1}^{g+1}}{\eta_0}.$$

Given assumptions for the values of the parameters η_0, μ^{g+1}, and $\overline{p}_{t+1}^{g+1}$, along with our DRM parameter estimates, we can solve for e_t and then simulate how the average level of effort among service members differs under TIS pay table versus the TIG pay table.

Modeling the Effect of Effort in Multiple Periods to Promote to the Next Grade

In the formulation above, the individual has some probability of being promoted in each period t, and the probability of promotion is dependent on effort in the immediately preceding period. In our model, as we described earlier in the chapter, we assume that the probability of promotion to a given grade occurs at a given number of YOS but that the probability of promotion differs by grade. That is, in our model promotion occurs at a given point in time for a particular grade. An implication of this approach to modeling promotion is that individual's promotion chances may depend on effort over multiple periods. We accommodate this feature

---

[9]   The derivation of this expression requires several steps. Appendix B shows these steps.

by changing the assumed form of the probability of promotion function. Instead of the probability being dependent on effort in a single period as follows:

$$p_{t+1}^{g+1} = \mu^{g+1} \overline{p}_{t+1}^{g+1} e_t$$

it can depend on effort in multiple periods, as in this example:

$$p_{t+1}^{g+1} = \mu^{g+1} \overline{p}_{t+1}^{g+1} \sum_{i=t-k}^{t} e_i.$$

The expressions for e_{t-1}, e_{t-2}, etc., take on a similar form to the expression for e_t. For example, the expression for e_{t-1} is

$$e_{t-1} = \frac{\beta^2 \Pr\left(\overline{V}^A\left(k_t\right) > V^L\left(k_t\right)\right) \Pr\left(\overline{V}^S\left(k_{t+1}\right) > V^L\left(k_{t+1}\right)\right) \left(\overline{V}^{A(g+1)}\left(k_{t+1}\right) - \overline{V}^{Ag}\left(k_{t+1}\right)\right) \mu^{g+1} \overline{p}_{t+1}^{g+1}}{\eta_0}.$$

Note that the values of $V^A(k_t)$ and $V^A(k_{t+1})$ depend on the value of e_t, e_{t+1}, e_{t+2}, and so on, so we cannot compute the value of e_{t-1} without knowing all the future levels of effort, as well as any past levels of effort associated with the same promotion point e_{t-1} is associated with. In general, if a promotion point probability depends on multiple years of effort, we need to solve for all the levels of effort associated with a promotion point simultaneously. So in our simulations we use an iterative procedure to solve for a set of levels of effort that are stationary; that is, we start off with a guess of the optimal level of effort in each period, and then solve for the optimal level of effort in each period given that all others are fixed, update the levels of effort, and iterate until the computed levels of effort cease to change. We solve for the levels of effort associated with the senior-most promotion point first, then the levels of effort associated with the next-most-senior promotion point, and so on until we work our way backward to the initial promotion point.

Solving for the optimal effort supply decision in each YOS for each member in our simulations is a nontrivial task. In the model, these decisions depend on only two parameters: the disutility of effort parameter and the relationship between promotion and effort. As with the ability parameters, we calibrated the effort-related parameters so as to replicate the cumulative retention profile. Figure 3.12 shows the fit for the Army enlisted model after calibrating the effort-related parameters where we ignore ability in the model. The simulated profile broadly tracks the observed profile, but the fit is not as good as the one in which we calibrate only the ability parameter, as shown in Figure 3.10. Consequently, in our presentation of results related to the effects of the TIG pay table on effort in the next chapter, we only show results for Army enlisted personnel and consider our results as exploratory.

Figure 3.12
Calibrating the Parameters of the Effort Decision, Army Enlisted Personnel

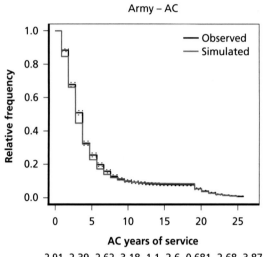

2.91, 2.39, 2.62, 3.18, 1.1, 2.6, 0.681, 2.68, 3.87

SOURCE: Authors' computations.

Summary

The DRM is a model with a relatively simple structure, but despite the simple structure it can support a rich variety of analyses. In this chapter, we extended it to model the promotion process and presented new estimates and model fits for enlisted personnel and officers for each service. We also extended the simulation capability to permit analysis of the TIG pay table and incorporated ability and the effort supply decision.

CHAPTER FOUR

Simulated Effects of a Time-in-Grade Pay Table on Retention, Performance, and Cost

This chapter presents the simulation results on the steady-state effect of a TIG versus a TIS pay table on retention over a career, performance and cost. Performance is measured in terms of promotion speed relative to peers, where we consider two factors that can affect performance: ability and effort supply. By *ability*, we mean characteristics of individual members that increase or decrease their promotion speed relative to their peers and can include innate cognitive intelligence as well as other characteristics that lead to success, such as ability to work well in teams and work in a hierarchical organizational structure and resilience to changes such as frequent moves and new assignments. By *effort supply*, or simply *effort*, we refer to how hard and effectively members work in terms of achieving tasks that lead to faster promotion. In terms of simulation, ideally, we would consider both ability and effort simultaneously as factors affecting promotion speed. As explained in more detail in Chapter Three, we consider them separately and incorporate ability into the DRM by making assumptions about

1. the extent to which ability differs among military entrants
2. the extent to which ability affects promotion speed
3. the effect of ability on external civilian opportunities.

We also conduct sensitivity analyses to assess how sensitive our results are with respect to these three assumptions. With regard to effort, we assume in the DRM that a member's promotion performance can depend on the amount of effort they exert. The main idea is that, other things held constant, the more effort a member exerts, the more likely they will be promoted. The structure of the model allows us to derive the optimal amount of effort an individual would exert given assumptions about how effort affects the probability of an individual being promoted, and assumptions about the disutility of effort. As might be expected, the optimal amount of effort is the level where the expected marginal return to effort equals the marginal disutility of effort. In this chapter, we first show the results related to ability and then to effort supply, with the latter analysis being more exploratory. In addition, we present simulated results of the effect of a TIG versus a TIS pay table on retention and cost. However, before presenting our simulation results, we first posit the results we might expect conceptually.

Conceptual Framework: How the Time-in-Grade Versus Time-in-Service Pay Table Might Affect Retention, Performance, and Effort Supply

Chapter Two showed that the TIG pay table provides a permanent reward and therefore greater lifetime compensation associated with faster promotion. To the extent that better performers are promoted more quickly, we would expect, conceptually, that the TIG table would have the following effects on retention, performance, and cost of the force:

- *Increased retention incentives for better performers, and reduced retention incentives for poorer performers:* The overall effect on retention is unclear and depends on the strength of the retention effects of better versus worse performers. If those who perform better have a stronger retention effect, we would expect overall retention to increase. Otherwise, we would expect it to decrease. If they are completely offsetting, we would expect overall retention to change little or not at all.
- *Increased average performance as measured by ability, across the force:* If higher-ability personnel are more likely to stay in service and lower-ability personnel are less likely to stay, we would expect average performance across the force to increase.
- *Ambiguous personnel cost per member:* If better performers are a larger share of the force, and compensation is higher for better performers under the TIG pay table, personnel costs per member will be higher under the TIG table. Cost per member would also increase if the force becomes more experienced under the TIG pay table. This could occur if the higher retention of better performers more than offset the lower retention of poorer performers. On the other hand, if the force becomes less experienced under the TIG table, cost per member could decrease or stay the same.[1]
- *Increased performance, on average, among those in higher grades:* To the extent that better performers are more likely to be promoted and retained, we would expect the average performance of those promoted and, therefore, in higher grades, to be greater under the TIG pay table.

In the case of ability as a metric of performance, we can also posit how the TIS versus a TIG pay table might affect the sorting of higher-ability personnel to higher grades. As discussed in prior research (Asch and Warner, 1994a, 2001; Asch, 2019b), an important function of the military compensation system as a human resource tool is to induce higher-ability personnel to stay in service and seek advancement to the upper grades. This is important because in a hierarchical organization such as the military, with virtually no lateral entry, the productivity of those in the upper ranks has spillover effects, either positive or negative, on the productivity of those in lower ranks. Given the hypothesis listed that we can expect increased performance on average among those in higher grades, we would expect the TIG to induce greater ability sorting, i.e., even higher ability on average in the upper grades than might exist under the current pay table.

---

[1]   When measuring costs per member, we hold total strength constant, thereby allowing us to focus on how changes in the experience mix of the force under the TIG pay table affects cost. However, by holding strength constant, we ignore the possibility that a more experienced and higher-ability force under the TIG pay table might allow the services to reduce strength. That is, they might be able to achieve the same level of readiness with a smaller force. As a result, total compensation costs could fall. We explore this point further when we conduct sensitivity analysis later in this chapter.

The next subsection shows simulation results of the effects on retention, ability, and cost of the TIG versus the TIS pay table. We then show results where performance is measured in terms of effort supply.

Simulated Effects on Retention and Ability Sorting of the Time-in-Grade Versus a Time-in-Service Pay Table

Figures 4.1 and 4.2 show simulated cumulative retention profiles under the TIS versus a TIG pay table for enlisted personnel and officers, by service, respectively.[2] The black and red lines are the simulated retention profiles under the TIS table and TIG table, respectively.[3] For enlisted personnel, we find that retention increases under the TIG table in each service primarily in the mid-career, though the Marine Corps shows the smallest increase. This implies that the positive effect of retention for those who are promoted faster more than offsets the negative effect on those who are promoted slower under the TIG table. For officers, we find almost no effect or a small negative effect across the services, implying that the positive and negative effects are about equal, with the negative effect stronger in some cases.

To quantify the retention effects, Table 4.1 and 4.2 show summary statistics of the effects of the TIG table relative to the TIS table, by service, for enlisted personnel and officers, respectively. With respect to retention, the tables show the percentage change in overall force size that we simulate under the TIG table compared with the TIS table. For enlisted personnel, the increase in force size ranges from 0.4 percent for the Marine Corps to 1.5 percent for the Army. For officers, the change in force size varied from –0.2 percent for the Army to 0.7 percent for the Marine Corps. The smaller effects for officers than enlisted could be due to the smaller effects of the TIG versus the TIS pay table for fast-promoting officers, due to the compression of the pay table discussed in the previous chapter in the context of Figures 2.3 and 2.4. An additional explanation is higher retention rates among officers than enlisted personnel, reflecting a relatively higher taste for service among officers than enlisted personnel. When taste or the persistent nonmonetary aspects for service is perceived as higher, personnel are relatively less responsive to changes in the monetary changes associated with staying in the military. The tables also show personnel costs per member in terms of basic pay and allowances and retirement accrual costs under a TIG versus the TIS pay table. In general, we find that the change in cost per member is relatively small, at most a 1 percentage point change, and is negative, except for Air Force officers.

The tables also summarize the simulated effects of the TIG table on performance as measured by ability percentile. We assume a normal distribution of ability at entry with mean 0. In percentile terms, the mean would be the 50th percentile of the distribution. We simulate

---

[2]   As a reminder, we consider ability and effort supply separately. Figures 4.1 and 4.2 show results incorporating ability, but not effort.

[3]   A brief note on interpreting the figures for readers who may have skipped Chapter Three, or who may wish to refresh their memory: The horizontal axis counts years since the individual was observed beginning active service. The vertical axis shows the cumulative probability of retention on active duty until that year. For example, at entry, YOS is 0 and the fraction of personnel retained is 1, and the fraction of the force retained falls over an active career as members leave active duty. The solid black line shows the actual retention of individuals in our cohorts, and the red line shows the predicted retention.

Figure 4.1
Enlisted Retention Under Time-in-Grade and Time-in-Service Pay Tables

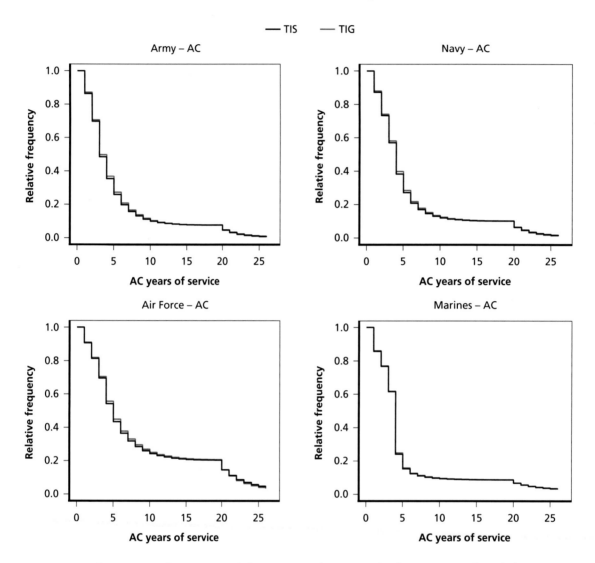

retention and compute the average ability percentile across the force retained and the average ability percentile at each grade. The tables show the average ability percentile across the force for each service, for enlisted personnel and officers, respectively, as well as average ability of personnel in E-5 and E-9 for enlisted and O-4 and O-7 for officers. The latter statistics indicate the extent of ability sorting: the retention and promotion of higher-ability personnel to the upper grades.

For enlisted personnel, we find that the average ability percentile across the force increases under the TIG pay table, but by less than 5 percent for any given service. For example, under the TIS pay table, the average ability percentile for Army enlisted personnel is 48.1, compared with 49.7 under the TIG pay table, an increase of 3.4 percent. We find no change for the Marine Corps, equal to 50.3 under both the TIG and TIS pay tables. The relatively small change of less than 5 percent for any service is not entirely unexpected, given past research (Asch, Romley, and Totten, 2005) on the retention and overall quality of the enlisted force using AFQT as the metric of personnel quality. In particular, research has found that the

Figure 4.2
Officer Retention Under Time-in-Grade and Time-in-Service Pay Tables

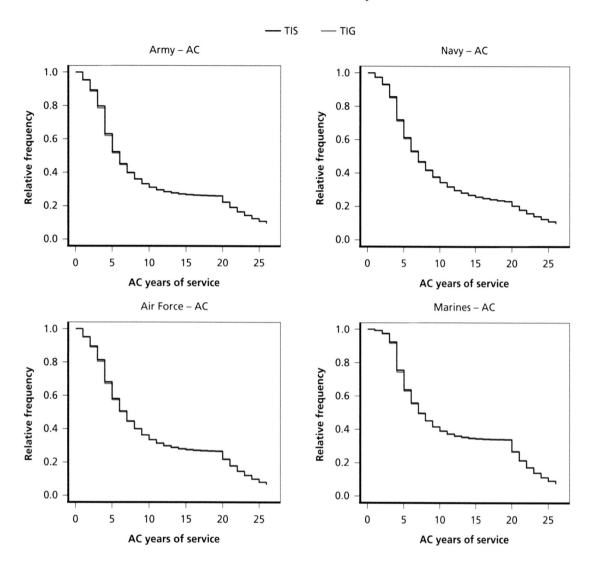

effects of better external opportunities for higher-quality enlisted personnel are generally offset by their better internal opportunities; the net result is that the quality of those who stay is not much different than the quality of those who leave. The main conclusion from this research is that the military's (TIS-based) compensation system is not strongly pro-selective on personnel quality. The simulations suggest that the TIG pay table also demonstrates relatively weak pro-selection, but importantly the pro-selection effect is nonetheless larger under the TIG than the TIS pay table for enlisted personnel.

We also find that both the TIS and TIG pay tables induce ability sorting for enlisted personnel, with the TIG pay table producing a strong effect. For example, the average ability percentile of an E-9 in the Army is 66.0, compared with 42.8 for an E-5 under the TIS pay table, an increase of 54.2 percent. In other words, enlisted personnel in the Army and in the other services promote and retain higher-ability personnel, resulting in higher average ability among those in the upper ranks under the current TIS pay table. It is notable that this result is consistent with earlier research using other metrics of personnel quality, such as AFQT, that

Table 4.1
Enlisted Summary Statistics by Service on Retention, Ability Sorting, and Cost

Enlisted Personnel	TIS Pay Table	TIG Pay Table
Army		
Average ability percentile		
E-5	42.8	43.6
E-9	66.0	76.9
Overall	47.3	48.9
Retention: percentage change in force size	0.0	1.5
Cost per member (2019 dollars)	$64,324	$64,173
Navy		
Average ability percentile		
E-5	44.4	44.8
E-9	69.5	76.6
Overall	48.6	49.5
Retention: percentage change in force size	0.0	1.3
Cost per member (2019 dollars)	$66,770	$66,582
Marine Corps		
Average ability percentile		
E-5	46.0	45.9
E-9	72.6	74.6
Overall	50.3	50.3
Retention: percentage change in force size	0.0	0.4
Cost per member (2019 dollars)	$65,105	$64,994
Air Force		
Average ability percentile		
E-5	43.0	43.4
E-9	65.8	71.4
Overall	47.1	48.1
Retention: percentage change in force size	0.0	1.2
Cost per member (2019 dollars)	$73,518	$73,244

SOURCE: Authors' computations.
NOTE: Costs include active duty basic pay and allowances and retirement accrual costs.

Table 4.2
Officer Summary Statistics by Service on Retention, Ability Sorting, and Cost

Officers	TIS Pay Table	TIG Pay Table
Army		
Average ability percentile		
O-3	31.1	31.3
O-7	72.6	75.7
Overall	36.6	37.3
Retention: percentage change in force size	0.0	−0.2
Cost per member (2019 dollars)	$123,989	$122,876
Navy		
Average ability percentile		
O-3	34.6	34.8
O-7	77.1	79.1
Overall	39.7	40.4
Retention: percentage change in force size	0.0	−0.3
Cost per member (2019 dollars)	$120,528	$119,331
Marine Corps		
Average ability percentile		
O-3	30.8	31.0
O-7	72.1	76.3
Overall	35.3	36.3
Retention: percentage change in force size	0.0	0.7
Cost per member (2019 dollars)	$127,814	$127,054
Air Force		
Average ability percentile		
O-3	31.0	31.1
O-7	74.9	77.0
Overall	36.1	36.9
Retention: percentage change in force size	0.0	0.1
Cost per member (2019 dollars)	$124,322	$123,401

SOURCE: Authors' computations.

NOTE: Costs include active duty basic pay and allowances and retirement accrual costs.

shows that the average quality of those in the upper enlisted ranks exceeds that in the lower ranks (Asch, Romley, and Totten, 2005). The key result, however, is that this effect is stronger under the TIG pay table. In particular, we find that the average ability percentile increases 76.3 percent (from 43.6 to 76.9) for the Army under the TIG pay table. This result occurs because better performers are more likely to be promoted and retained under the TIG pay table. We find similar results for enlisted personnel in the other services.

For officers, Table 4.2 shows that the average overall ability percentile is also higher under the TIG pay table than the TIS pay table. As with enlisted personnel, the percentage change is less than 5 percent for any given service. For example, for the Army, the average increases from 36.6 to 37.3, an increase of 1.9 percent. We also find improved ability sorting under the TIG pay table for officers. While the simulations show the average ability percentile is higher among O-7s than O-3s for any given service under both the TIS and TIG pay tables, the difference is greater under the TIG pay table, though the amount varies across the services.

Efficiency

A key result of our simulations for enlisted personnel above is that retention increases under the TIG pay table versus the TIS pay table, with virtually no change in cost per member. This result implies that the TIG pay table is more efficient—more readiness is produced by the TIG pay table for the same cost. An additional implication is that about the same retention could be achieved under the TIG pay table with less cost. We illustrate this implication in Table 4.3 using Army enlisted personnel as an example and consider as an example a 0.375 percent pay cut as a means of reducing force size. We show that a 0.375 percent across-the-board pay cut under the TIG pay table would lead to force size equivalent to force size under the TIS pay table. Although force size is the same, cost per member is lower, $63,634 versus $64,173. Furthermore, the TIG pay table, even with an across-the-board pay cut, still results in stronger ability sorting than the TIS pay table. The results imply that the TIG pay table would enable DoD to achieve existing readiness objectives related to retention and increase ability sorting at the same cost per member.

Table 4.3
Army Enlisted Summary Statistics with 0.375 Percent Across-the-Board Pay Cut Under the Time-in-Grade Pay Table

Army Enlisted Personnel	TIS Pay Table	TIG Pay Table	TIG Pay Table with 0.375% Across-the-Board Pay Cut
Average ability percentile			
E-5	42.8	43.6	43.7
E-9	66.0	76.9	76.8
Overall	47.3	48.9	48.9
Retention: percentage change in force size	0.0	1.5	0.0
Cost (2019 dollars)	$64,324	$64,173	$63,634

SOURCE: Authors' computations.
NOTE: Costs include active duty basic pay and allowances and retirement accrual costs.

Sensitivity Analyses

As we discussed in Chapter Three, we make three assumptions about ability to incorporate it into our DRM simulations:

1. the effect of ability on promotion timing
2. the effect of ability on external opportunities
3. the extent to which ability varies among military entrants.

The specific assumptions we make are ones that allow us to replicate the steady-state retention profiles of enlisted personnel and officers in each service under the TIS pay table.

This subsection shows sensitivity analyses to assess the extent to which our main conclusions about the retention, performance, and cost effects of the TIG pay table change under alternative assumptions. We conduct sensitivity analysis in which we vary each of these assumptions using our DRM model for Army enlisted personnel. In particular, we consider the following three sensitivity analyses:

1. Increase the responsiveness of external opportunities to differences in ability. In the main analysis, we assume that external civilian basic pay are proportionate to the standard deviation of ability according to the formula $(1 + 0.1 \times \sigma_a)$, where σ_a is the standard deviation of the ability distribution. In the sensitivity analyses, we assume a formula of $(1 + 0.2 \times \sigma_a)$.
2. Reduce the responsiveness of promotion speed to differences in ability. In the main analysis, we assume that promotion time to E-6 and above varies in proportion to ability by one year. In our sensitivity analyses, we assume that promotion time to E-7 and above varies in proportion to ability by one year.
3. Reduce the variation in ability among entrants. As discussed in Chapter Three, for enlisted personnel, we assume a standard deviation of the ability distribution of 0.5. For the sensitivity analyses, we reduce it to 0.25.

We report the results of these sensitivity analyses in Table 4.4. Specifically, the table shows summary statistics for Army enlisted personnel under the TIS and TIG pay tables for each of the three analyses. Our results remain qualitatively the same under each of the three analyses. In particular, as in the main analyses, we find that retention increases under the TIG pay table relative to the TIS pay table. Furthermore, we find that cost per member falls slightly across the three analyses, by less than 1 percent, similar to the main analysis. We also find that it is still the case that ability sorting improves under the TIG pay table. Finally, we find that the overall quality of the force increases in each case.

Exploratory Analysis: Simulated Effects on Effort Supply of the Time-in-Grade Versus Time-in-Service Pay Table

Separate from ability, we also simulated the retention, cost, and performance effects of the TIG pay table when performance is measured in terms of effort supply. As we explained in Chapter Three, we assume parameters of the effort supply decision such that we can replicate the observed retention profile under the TIS pay table. The assumed parameters are the disutility

Table 4.4
Army Enlisted Summary Statistics: Sensitivity Analyses

Army Enlisted Personnel	1. Increase the Effect of Ability on External Opportunities		2. Reduce the Effect of Ability on Promotion Timing		3. Reduce Variability in Ability Among Entrants	
	TIS Pay Table	TIG Pay Table	TIS Pay Table	TIG Pay Table	TIS Pay Table	TIG Pay Table
Average ability percentile						
E-5	35.9	36.7	44.4	44.7	41.4	41.9
E-9	48.3	60.5	54.1	62.3	54.4	62.6
Overall	40.9	42.3	45.6	46.3	43.7	44.6
Retention: percentage change in force size	0.0	0.6	0.0	1.0	0.0	1.0
Cost (2019 dollars)	$65,385	$64,786	$64,576	$64,117	$64,107	$63,779

SOURCE: Authors' computations.

NOTE: Costs include active duty basic pay and allowances and retirement accrual costs.

of effort parameter and the parameter representing the relationship between effort and promotion to each grade. As we showed in the earlier chapter, we were moderately successful in replicating the observed retention profile. Consequently, our simulation results regarding effort supply should be considered more suggestive than the results shown above, where performance is measured in terms of ability.

Table 4.5 shows summary statistics for Army enlisted personnel. As a reminder, the level of effort directly influences the probability of promotion; however, this effort does not come without cost to the individual, because the associated disutility in a period is assumed to go up as the square of the level of effort in that period. We compute the optimal level of effort for all individuals in every time period they are retained and look at the average level of effort over the force to gauge the effect of the different pay tables on effort. The key result is that average effort across the force increases under the TIG pay table relative to the TIS pay table, 0.97 versus 0.89, or an overall increase in effort supply of 9 percent. As hypothesized above, given that the financial rewards to promotion are greater under the TIG pay table, the financial incentives to increasing effort supply are higher, insofar as better performers are promoted faster. Muting this effect is the disutility associated with increased effort. Overall, we find that average effort across the force is higher under the TIG versus the TIS pay table in the Army enlisted personnel example. We also find that overall force size increases by 4.4 percent, while the cost per member increases by only 0.6 percent. The TIG pay table increases retention and performance, when performance is measured in terms of effort supply.

Summary

The key result of the simulations shown this chapter is that the TIG pay table would be a more efficient approach to setting basic pay. For enlisted personnel, we find that simulated retention in the steady state would increase under the TIG pay table, while personnel costs per member would be generally fall, albeit by at most 1 percent. For officers, retention in the steady state as

Table 4.5
Army Enlisted Summary Statistics Using Effort as the Metric of Performance

Army Enlisted Personnel	TIS Pay Table	TIG Pay Table
Average effort	0.89	0.97
Retention: percentage change in force size	0.0	4.4
Cost (2019 dollars)	$65,631	$66,019

SOURCE: Authors' computations.
NOTE: Costs include active duty basic pay and allowances and retirement accrual costs.

well as cost per member would change little (either positive or negative). On the other hand, the simulations indicate that performance would increase overall across the force and in the upper grades relative to the lower grades. We demonstrated, using the Army enlisted force as an example, that greater performance could be achieved at less cost and for the same retention under the TIG table relative to the TIS pay table. We also conducted sensitivity analyses in which we altered the underlying assumptions of our simulations with respect to ability, and we found that the results generally remain unchanged.

CHAPTER FIVE

Transition Costs and Save Pay

The previous chapter focused on steady-state effects when all members have spent an entire career under the TIG pay table. But, another area of concern is the effect of the TIG pay table during the transition period. Commissions as early as the 1957 Defense Advisory Committee on Professional and Technical Compensation raised the concern that members would see a reduction in pay during the transition from the TIS to the TIG pay table. Like later commissions, including the DACMC, the 1957 commission recommended "save pay," a policy that would prevent members from receiving lower compensation than before the change. In the case of the Defense Advisory Committee on Professional and Technical Compensation, it specifically recommended that pay be frozen at its present level until the member qualifies for promotion. In this chapter, we consider the transition effects of the TIG pay table from the standpoint of the effects on members' basic pay before and after the transition. First, we estimate the share of active duty members that would experience either a pay increase or decrease in the first year of the TIG basic pay policy and the extent of the pay increase or decrease. We then estimate the first-year cost of a save pay policy that would ensure members would receive at least the same amount of basic pay under the TIG pay table as they did under the TIS pay table.

Transition Effects on Member Pay

Reductions or increases in basic pay for a given member can occur after the transition to the TIG pay table because of the way the TIG table is constructed. As we described in Chapter Two, anchor points or entry YOS for the construction of the TIG pay table were chosen based on average promotion times observed between FY 2013 and 2018. For example, the entry YOS for E-6 is 6, meaning that basic pay in the TIG pay table for a member recently promoted to E-6 with 0 time in grade is equivalent to that of an E-6 with 6 YOS in the TIS pay table. An implication of the choice of entry YOS anchor points is that basic pay may be higher or lower for a given member in the year of transition to the TIG pay table if the member's promotion timing to a given grade deviates from the assumed entry YOS for that grade. As we'll discuss more in Chapter Seven, promotion times for individual service members can vary considerably from the averages shown in Table 2.1. Consequently, promotion times do differ from the entry YOS anchor points used to construct the TIG pay table.

For example, an E-6 with 12 YOS and 6 years in grade as an E-6 at the time of transition would receive the same pay after the transition to the TIG pay table as before the transition. The reason is that this E-6 was promoted to E-6 at 6 YOS (12 − 6 years) which is the same

YOS as we assume for the entry YOS anchor point. Consequently, the pay of this member is the same in both the TIS and TIG pay table.

But, instead, if the E-6 with 12 YOS had, say, 2 years in grade at the time of transition, the member's monthly basic pay would decrease in the transition year, from $3,776.70, the pay of an E-6 with 12 YOS in the TIS pay table to $3,453.60, the pay of an E-6 with 2 years in grade. The reason is that this E-6 was promoted to E-6 at 10 YOS (12-2 years), or at a YOS that is greater than the assumed entry YOS to E-6. Similarly, if the E-6 with 12 YOS had, say 7 years in grade, the member's monthly basic pay would increase instead, from $3,046.20 to $3,776.70. The reason is that the member's years of service at promotion to E-6 (12-7) were less than the assumed entry YOS anchor point of 6 for an E-6.

We investigated the extent to which members on active duty would experience an increase or decrease in pay using DMDC active duty master file data for all active duty members in service in January 2019 together with the 2018 TIS basic pay and associated TIG pay table in Table 2.2. The DMDC data provided information on the time in current grade, time in grade and YOS at promotion, and YOS for each member on active duty. Table 5.1 shows tabulations of the percent of personnel who would receive the same basic pay in the year of transition, lower pay in the TIG pay table, or higher pay in the TIG pay table for enlisted personnel, commissioned officers, warrant officers, and officers transitioning from enlisted service in grades O-1E to O-3E. Across all active duty personnel, 45.7 percent would receive the same basic pay, about one-third (32.1 percent) would experience a pay reduction as a result of the transition to the TIG pay table, and 22.3 percent would experience a pay increase.

The percentages differ by grade category. Nearly all warrant officers (91.6 percent) would experience a pay reduction, while about half (or 53.2 percent) of commissioned officers would experience a pay reduction in the transition to the TIG pay table. On the other hand, the majority of enlisted personnel who became officers and are in pay grades O-1E to O-3E would experience a pay increase. In the case of enlisted personnel, about one quarter (27.1 percent) would experience a pay reduction.

Table 5.1 also shows that among the 32.1 percent of members who would experience a pay reduction, the reduction in basic pay would average 6 percent. The extent of the reduction

Table 5.1
Extent of the Change in Basic Pay in the Year of Transition to the Time-in-Grade Pay Table from a Time-in-Service Pay Table

	Percentage of Members			Given Pay is Lower in TIG Table
	Same	Lower in TIG Table	Higher in TIG Table	Average Percentage Difference in Basic Pay
Enlisted	50.2	27.1	22.7	−5.2%
Commissioned officers	29.3	53.2	17.5	−6.6%
O-1E to O-3E	2.6	44.2	53.2	−8.5%
Warrant officers	3.1	91.6	5.4	−15.0%
All	45.7	32.1	22.3	−6.0%

SOURCE: Authors' computations.

NOTE: Tabulations based on the 2018 TIS and TIG pay tables (see Tables 2.2 and A.1) and DMDC data on active duty members in January 2019.

varies with grade category. Basic pay would decrease for 91.6 percent of warrant officers, and the average reduction in monthly basic pay would be 15.0 percent. The average reduction for enlisted personnel would be 5.2 percent and commissioned officers would be 6.6 percent; it would be 8.5 percent for those in the grades of O-1E to O-3E.

In short, based on the promotion histories of members on active duty in January 2019, we find that a sizable segment of the force would experience a reduction in pay at the time of transition. Furthermore, the reduction for these members is sizable; the last time basic pay changed by more than 6 percent in any given year (in absolute value) was in 1986 (DoD, 2018).

Save Pay

Save pay refers to a policy that "saves" an individual's rate of pay in situations in which a change in position or other policy causes an individual to be entitled to a lower rate of pay than before the change (DoD, Under Secretary of Defense [Comptroller], 2017; Office of Personnel Management, undated). Save pay is a policy that is already being used by DoD for both uniformed and civil service personnel. In the case of military personnel, enlisted personnel who accept an appointment as an officer and face a reduction in pay as a result of that transition can receive save pay in the form of the pay that they would have received in their last enlisted grade. Similarly, warrant officers who transition to commissioned officers can receive the pay they would have received in their last warrant officer grade or the pay in their last enlisted grade if they had previously been enlisted members (DoD, Under Secretary of Defense [Comptroller], 2017).

The DACMC and 10th QRMC estimated that the first-year cost of a save pay transition provision that held members "harmless" in terms of basic pay would be about $1.1 billion based on the 2005 pay table.[1] In 2018 dollars, this figure would be $1.43 billion. The 10th QRMC also considered a different save pay option instead of the "hold members harmless" provision. The alternative would ensure that there were no nominal reductions in the level of basic pay. If the transition to the TIG pay table occurred at the same time as the annual military pay raise, then part of the cost of the transition could be "covered" by the cost of the annual pay raise. Furthermore, if the post-transition basic pay under the TIG pay table also allowed for any pay raise associated with promotion occurring in the first year, then save pay costs would be further reduced, since part of the cost of the transition could also be "covered" by the cost associated with promotion-related pay raises. Under this save pay approach, the 10th QRMC estimated that the cost would be about $354 million rather than $1.1 billion, or about a third of the cost.

Following the "hold members harmless" approach, we estimated the first-year cost of save pay using the January 2019 data on the active force. Table 5.2 shows the results. We find that the first-year transition-cost across the active force would be $1.39 billion in 2018 dollars. Most of the cost is associated with the enlisted force ($0.61 billion). The $1.39 billion figure is very close to the $1.43 billion estimated by the Hogan and Mackin (2008) for the 10th QRMC, in 2018 dollars. To put the $1.39 billion figure in context, the 2018 appropriation for active component military personnel was about $115.9 billion (DoD, 2019).[2]

---

[1]   The 10th QRMC approach is assessed and discussed in Hogan and Mackin (2008).

[2]   This figure excludes Medicare-Retiree Health Care Contributions.

Table 5.2
Cost of Save Pay in the Year of Transition to the Time-in-Grade Pay Table from a Time-in-Service Pay Table (2018 dollars, billions)

	Cost (billions of dollars)
Enlisted	0.61
Commissioned officers	0.54
O-1E to O-3E	0.07
Warrant officers	0.17
All	1.39

SOURCE: Authors' computations.

NOTE: Tabulations based on the 2018 TIS and TIG pay tables (see Tables 2.2 and A.1) and DMDC data on active duty members in January 2019.

We do not estimate save pay costs under an approach that holds pay at nominal levels like the 10th QRMC did. But, as a rough order of a magnitude, if we use the 10th QRMC estimate that cost would be about a third, we would estimate a cost of about $460 million.

Summary

To the extent that the promotion times of service members vary from the average promotion times that were used to construct the TIG pay table, service members will experience an increase or decrease in their monthly basic pay at the time of transition to the TIG pay table from the TIS table. Based on the number of YOS and promotion history of active duty personnel in service in January 2019, we estimate that about one-third would experience a basic pay reduction, or 32.1 percent. We estimate that 45.7 percent would receive the same basic pay and 22.3 percent would experience a pay increase. We estimate that the average reduction would be 6 percent among those who would experience a reduction in pay at the time of the transition. If DoD adopts a policy to hold members harmless in terms of the level of basic pay by offering save pay, we estimate that in the first year, the cost of this save pay policy would be $1.39 billion in 2018 dollars, with most of the cost being attributed to save pay for enlisted personnel.

CHAPTER SIX

Two Alternative Performance-Based Policies Under a Time-in-Service Pay Table

The 13th QRMC requested that RAND investigate alternative approaches to reward better performance other than a TIG pay table that could be implemented under the current TIS pay table. Specifically, it requested an exploration of two concepts: constructive credit for performance and credential pay or pay to members who earn a specific skill credential. While DoD already has a constructive credit policy, it is not currently structured to reward superior performance of military members already in service. DoD also has credential pay, called skill incentive pay in Section 353 of Title 37 of the U.S. Code. Under this section, the services have the authority to offer skill incentive pay, a monthly amount that can be paid to service members who serve in a career field or skill designated as critical by the service secretary. However, skill incentive pay is not structured to be a pay-for-performance mechanism. We summarize our analysis and findings of these two concepts in this chapter.

Constructive Credit for Faster Promotion

Constructive credit, as currently implemented by DoD, rewards service members for advanced education, training, or experience earned prior to entering the military. The policy gives YOS credit for these activities, thereby allowing these individuals to enter service at a higher starting grade and, consequently, at higher military basic pay than they would in the absence of constructive credit. The use of constructive credit is limited to occupations in the medical field, legal field, and chaplains, though, for a short period from 2014 to 2018, constructive credit could also be applied to those with a background in cyber. As discussed in previous chapters, as a result of expanded constructive credit authority included in the 2019 NDAA, officers can enter service at a grade as high as O-6.

Current policy regarding constructive credit focuses on providing higher entry pay for lateral entrants than they would receive if they entered as an O-1. In Chapter Two, we noted that under the TIG pay table, lateral entrants would receive higher pay than under the TIS pay table because in the latter case, entrants would still enter with 0 YOS. We also showed that if the concept of constructive credit were expanded to also give service members YOS credit in the pay table, pay under the TIS pay table could be equivalent to pay under the TIS pay table for lateral entrants. Thus, it is possible to achieve the same pay outcome under the TIS pay table for lateral entrants.

In this chapter, we consider a further expansion of the definition of constructive credit that would give YOS credit in the pay table for better performance. In particular, we consider a policy that would give personnel who are promoted faster than their peers a permanent 1 YOS leg up in the pay table for the purpose of computing basic pay. The purpose of the policy would be to provide a permanent reward for fast promotion, something that is missing under the TIS pay table. Note, however, that constructive credit for performance would affect longevity for the purpose of computing a member's basic pay, but not for the purpose of retirement eligibility for computing retired pay.

For example, suppose a member is promoted to O-4 one year ahead of their peers, say at YOS 10 rather than YOS 11 like the rest of peer group. Under current policy, this member would receive the pay of an O-4 with 10 YOS, while this member's peers would receive the pay of an O-3 with 10 YOS. One year later, when the rest of the peer group is promoted, both the fast promote and the on-time promotes would receive the pay of an O-4 with 11 YOS. But, under an expanded definition of constructive credit that rewarded faster promotion, the member who was promoted faster would receive the pay of an O-4 with 11 YOS, and the member's on-time peers would receive the pay of an O-3 with 10 YOS. One year later, the fast promote would receive the pay of an O-4 with 12 YOS, and the member's on-time peers would receive the pay of an O-4 with 11 YOS. Thus, the constructive credit policy provides a permanent reward to the fast promotee who, in our example, is promoted one year ahead of their peers.

Effects of Constructive Credit for Performance on Basic Pay over a Career

We illustrate how constructive credit for performance would affect basic pay over a career by considering the effects on enlisted personnel and officers. Figure 6.1 replicates Figure 2.4 by showing a comparison of pay over a career for an officer under the TIG versus the TIS pay table for an officer promoted a year early to O-4. In addition, Figure 6.1 shows pay over a career for an O-4 who is given constructive credit for performance. Similarly, Figure 6.2 replicates the left panel of Figure 2.5 by showing a comparison of pay over a career for a fast-promoting enlisted occupation (DoD Occupation Code 0) under the TIG versus the TIS pay table. As mentioned in Chapter Two, occupations within DoD Occupation Code 0, Infantry, Gun Crews, and Seamanship Specialists, promote about one year faster to E-5 and E-6. Figure 6.2 also shows pay over a career for an enlisted member who receives constructive credit for performance. The pay profile under constructive credit is shown by the red line in the two figures.

We find that the basic pay profiles for fast promoters under the TIS pay table are higher with constructive credit than without constructive credit. That is, the red line is above the blue line for the TIS pay table without constructive credit. Furthermore, the higher pay profile under the TIS pay table with constructive credit is nearly identical to the TIG pay profile. The implication of this analysis is that constructive credit for performance is a policy that can broadly replicate the higher pay found under the TIG pay table.

Simulations of the Effects of Constructive Credit on Retention, Cost, and Ability

We next investigate whether constructive credit for performance can also broadly replicate the stronger incentives for performance and the increased efficiency of achieving retention and performance outcomes, as we found under the TIS pay table using Army personnel as an example. Figure 6.3 replicates results from Figure 4.1 for Army enlisted personnel and Figure 4.2 for Army officers but also shows simulated retention profiles under the TIS pay table with constructive credit for performance. Similarly, Table 6.1 replicates results from Tables 4.1 for

Figure 6.1
Simulated Monthly Basic Pay over a Career, Time-in-Grade Versus Time-in-Service Pay Tables Versus Time-in-Service Pay Table with Constructive Credit for Fast-Promoting Officers

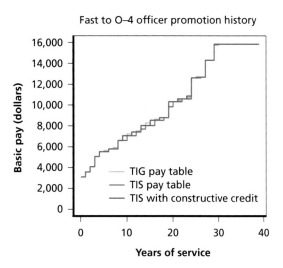

SOURCE: Authors' computations.

Figure 6.2
Simulated Monthly Basic Pay over a Career, Time-in-Grade Versus Time-in-Service Pay Tables Versus Time-in-Service Pay Table with Constructive Credit for Fast-Promoting Enlisted Occupation (DoD Occupation Code 0)

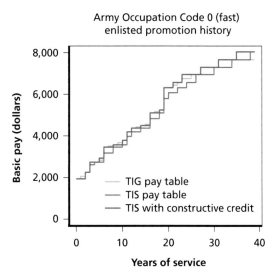

SOURCE: Authors' computations.

Army enlisted personnel and Table 4.2 for Army officers but also shows results under the TIS pay table with constructive credit for performance.

We find that, relative to retention under the TIS pay table, retention for Army enlisted personnel improves more under the TIG pay table than under a TIS pay table with constructive credit. As shown in Figure 6.1, the red line representing retention under the TIG pay table is higher in the mid-career while the green line representing retention under a TIS pay table

Figure 6.3
Army Enlisted and Officer Retention Under Time-in-Grade Versus Time-in-Service Pay Tables Versus Time-in-Service Pay Table with Constructive Credit

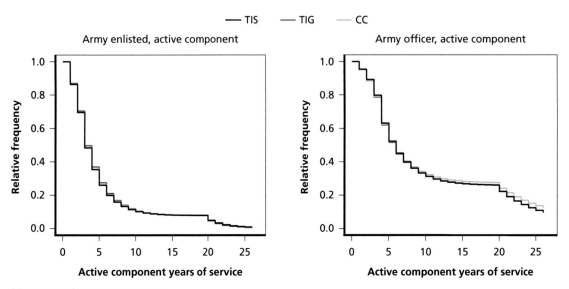

SOURCE: Authors' computations.
NOTE: CC in the legend refers to a TIS pay table with constructive credit for performance.

with constructive credit is slightly higher than retention under the TIG pay table for years beyond 10 YOS. As shown in Table 6.1, force size increases by 1.2 percent under a TIS pay table with constructive credit compared with 1.5 percent under the TIG pay table.

For Army officers, retention is higher under a TIS pay table with constructive credit than under either the TIS pay table alone or the TIG pay table, particularly later in the officer career as shown in the right panel of Figure 6.3. As shown in Table 6.1, the officer force size increases by 1 percent, compared with –0.2 percent under the TIG pay table.

Table 6.1 shows simulation results pertaining to the retention of higher-ability personnel and cost per member. For Army enlisted personnel, the average ability percentile increases under a TIS pay table with constructive credit relative to the TIS pay table without constructive credit, from 47.3 to 48.3, but does not increase as much as under the TIG pay table (48.9). Similarly, constructive credit for performance results in improved ability sorting relative to a TIS pay table without constructive credit with the average ability percentile for an E-9 increasing from 66.0 to 73.2. But the increase is not as large as under the TIG pay table, where the average ability percentile for an E-9 increases to 76.9 For Army officers, the average ability percentile is also lower, albeit slightly, under constructive credit versus the TIG pay table, though, as with enlisted personnel, it is higher than under a TIS pay table without constructive credit. On the other hand, ability sorting in terms of the difference between the average ability percentile of O-7 versus an O-3 is improved relative to both the TIS and TIG pay tables. However, this improvement is attributable to lower average ability of O-3s and not to higher ability of O-7s compared with either the TIG or TIS pay table, so the overall result cannot be viewed as a positive overall. In short, for both enlisted personnel and officers, average ability and ability sorting improve under a TIS pay table with constructive credit but not as much as under the TIG pay table.

Table 6.1
Army Enlisted and Officer Summary Statistics Under a Time-in-Service Pay Table with Constructive Credit

	TIS Pay Table	TIG Pay Table	TIS Pay Table with Constructive Credit
Army Enlisted Personnel			
Average ability percentile			
E-5	42.8	43.6	43.4
E-9	66.0	76.9	73.2
Overall	47.3	48.9	48.3
Retention: percentage change in force size	0.0	1.5	1.2
Cost (2019 dollars)	$64,324	$64,173	$64,748
Army Officers			
Average ability percentile			
O-3	31.1	31.3	28.1
O-7	72.6	75.7	75.8
Overall	36.6	37.3	37.1
Retention: percentage change in force size	0.0	−0.2	1.0
Cost (2019 dollars)	$123,989	$122,876	$124,503

SOURCE: Authors' computations.
NOTE: Costs include active duty basic pay and allowances and retirement accrual costs.

The results for enlisted personnel in Table 6.1 also show that constructive credit is less efficient than the TIG pay table. Cost per member is lower under a TIG pay table, $64,173 versus $64,748, and constructive credit improves retention by less, 1.2 percent versus 1.5 percent; improves average ability by less, 48.3 versus 48.9; and results in less ability sorting. That said, constructive credit is an improvement over the TIS pay table in terms of efficiency, at least in terms of ability sorting (Table 6.2). We find that constructive credit with a slight pay cut of 0.18 percent would result in the same retention and cost per member as a TIS pay table without constructive credit, but average ability across the force and among E-9s would be greater with constructive credit.

Skill-Based or Credential Pay

Credential pay refers to additional monthly compensation that a military service member could receive for holding a specific educational or training credential. Our investigation of credential focused on whether it could provide incentives for performance similar to what could be provided by the TIG pay table. Our approach involved reviewing the available academic and defense manpower literature on credential pay. We summarize our review of the literature in this subsection. We first review the different names and definitions used to describe credential pay in the literature and discuss the relevance to our investigation of the literature that focuses

Table 6.2
Army Enlisted Summary Statistics Under a Time-in-Service Pay Table with Constructive Credit and a 0.18 Percent Across-the-Board Pay Cut

Army Enlisted Personnel	TIS Pay Table	TIG Pay Table	TIS Pay Table with Constructive Credit	TIS Pay Table with Constructive Credit and 0.18 Percent Pay Cut
Average ability percentile				
E-5	42.8	43.6	43.4	43.4
E-9	66.0	76.9	73.2	73.1
Overall	47.3	48.9	48.3	48.4
Retention: percentage change in force size	0.0	1.5	1.2	0.0
Cost (2019 dollars)	$64,324	$64,173	$64,748	$64,318

SOURCE: Authors' computations.
NOTE: Costs include active duty basic pay and allowances and retirement accrual costs.

on the private sector. Next, we review the defense manpower literature including the report by Davis and Horowitz (2008) prepared for the 10th QRMC, which also considered credential pay as a performance-based pay alternative to a TIG pay table. We conclude with a summary of the advantages and disadvantages of credential pay based on our literature review.

Alternative Credential Pay Definitions and the Relevancy of the Nondefense Manpower Literature

Credential pay in the academic literature is also alternatively known as proficiency pay, certification pay, skill-based pay, skill pay, and knowledge-based pay, though the typical name used is skill-based pay. Under all of these definitions, the key concept is that pay is based on the skills an employee possesses. In the literature that focuses on the nonmilitary population, researchers frame credential or skill-based pay as an alternative pay setting approach that is based on job classification or the tasks and responsibilities associated with a given job. For example, Gupta, Jenkins, and Curington (1986) define skill-based pay as a compensation system that bases salaries and wage rates not on particular job classifications but on the skills and competencies an employee possesses. As described by Ledford and Heneman (2011, p. 1), "skill-based pay is a person-based system, because it is based on the characteristics of the person rather than the job. In more common job-based pay systems, pay is based on the job, which employees are entitled to receive even if they are not proficient in their position." An example of a job-based pay system is the General Schedule system for federal employees. General Schedule pay rates are based on job classifications. While the hiring and promotion of employees to different jobs may be based on the employees' skills and experiences, the General Schedule pay rate offered to an employee entering a given job is not higher if the employee possesses more skills. Papers that have evaluated skill-based pay rather than jobs-based pay using private or public sector data include Parent and Weber (1994), Guthrie (2000), Murray and Gerhart (1998), Luthans and Fox (1989), Mitra, Gupta, and Shaw (2011), and Lockey, Graham, and Zhou (2017).

The definition of skill-pay or credential pay differs in the military context. In the military, skill pay is a bonus or additional pay that is provided in addition to basic pay for demonstrated proficiencies. Ledford and Heneman (2011) note that the skill-based pay used in the military is unique and almost unknown outside the military. Because of the uniqueness of the military, the applicability of the nondefense manpower literature is limited, and we focused the rest of our review of the literature on military-related studies. Before summarizing the four studies we identified, we first provide an overview of the history of skill pay and proficiency pay in the military.

History of the Use of Proficiency Pay in Military

Between 1958 and 1985, the military services had authority to offer proficiency pay to qualified members (Davis and Horowitz, 2008; Hosek and Asch, 2002). The purpose of proficiency pay was to induce the retention of enlisted personnel who were required to perform "extremely demanding duties or duties demanding an unusual degree of responsibility" and to induce "qualified personnel to volunteer for such duties" (DoD, 1996, p. 477). Proficiency pay resulted from deliberations of the Defense Advisory Committee on Professional and Technical Compensation. In 1957, it recommended a change in the pay structure that would allow the promotion of a member to a higher pay grade without promotion to a higher rank. According to Hosek and Asch (2002), the intent of the committee was to create a pay for members who were specifically proficient in a given skill.

The Uniformed Services Pay Act of 1958 permitted the service secretaries to "choose such a proficiency pay grade" method "designated as . . . specially proficient in a military skill" (DoD, 1996, p. 477). It also permitted the service secretaries to alternatively pay a flat rate of up to $150 per month as proficiency pay. They chose the latter method and never used the proficiency pay grade method. Three types of proficiency pay were established: shortage specialty proficiency pay, special duty assignment proficiency pay, and superior performance proficiency pay. Shortage specialty proficiency pay was displaced by the selective reenlistment bonus in 1975 and phased out rapidly. By 1977, only 7,000 people were receiving shortage specialty pay, compared with 135,000 in 1975 (Hosek and Asch, 2002). In 1982, the shortage specialty pay program was absorbed into the special duty assignment pay program. Superior performance pay was authorized until 1976 and then terminated. According to Davis and Horowitz (2008), this pay failed largely because it was unpopular; singling out members for extra pay was unpleasant for defense managers. Special duty assignment proficiency pay was paid to personnel performing such voluntary duties as recruiters, drill instructors, or reenlistment noncommissioned officers. In 1985, new proficiency pay authority limited such pay to special duty assignments (the word *proficiency* was dropped).

Of the three pays, only superior performance proficiency pay is closely related to the intent of the 1957 Defense Advisory Committee on Professional and Technical Compensation. In practice, proficiency pay focused on increasing retention in specialties with shortages—a role taken over by the selective reenlistment bonus program. Proficiency pay for arduous assignment was also not related to a member's skill proficiency, and, not surprisingly, the special duty assignment pay program eliminated the term *proficiency*. In any case, the proficiency pay program, as ultimately used by the services between 1958 and 1985, did not provide a payment that intended to help the services create and preserve a stock of a particular skill.

In response the 2006 DACMC and the 10th QRMC, in 2008 Congress consolidated the 65 categories of special and incentive pays into eight general categories and gave DoD ten years to implement the consolidation. One of these broad categories is called "Skill Incentive-Proficiency Pay," mentioned as "skill incentive pay" above and created in 2016, according to the U.S. Government Accountability Office (2017). Special duty assignment pay was transitioned to 37 U.S.C. Section 352, "Assignment Pay or Special Duty Pay." the authority to pay special duty assignment pay under the old code (37 U.S.C. section 307) expired on January 27, 2018.

Since 2009, 37 U.S.C. Section 353 has defined "skill incentive pay or proficiency bonus." Section 353(a) defines skill incentive pay. The service secretaries may pay a monthly skill incentive pay to a member of a regular or reserve component of the uniformed services who is entitled to basic pay and who serves in a career field or skill designated by the service secretary. The maximum amount cannot exceed $1,000 per month. The amount can be prorated if an individual is not eligible for the entire month. Certification is required annually. A member can't be paid more than one pay under this section in any month for the same period of service and skill or be paid hazardous duty pay for the same period and for the same skill.

Section 353(b) defines skill proficiency bonus: The service secretaries may pay a proficiency bonus to a member of a regular or reserve component who is entitled to basic pay and is determined to have and maintains certified proficiency in a designated skill deemed critical by the service secretary or is in training to acquire proficiency in a critical foreign language or expertise in a foreign cultural studies or a related skill designated as critical. The bonus may be paid in lump sum at the beginning of the proficiency period or in periodic installments. The amount may not exceed $12,000 for each 12-month period. Military Foreign Language Skill Proficiency Bonuses is an example of a skill proficiency bonus.

Finally, although not explicitly considered proficiency pay, the services offer reenlistment bonuses that are targeted to specific occupations and, in some cases, to specific skill areas within an occupation. These areas can represent an advanced skill that is not held by all members in the occupation. For example, an Army combat medic who reenlists and also has an additional skill identifier indicating the individual is a nationally registered flight paramedic might receive a higher reenlistment bonus than other combat medics. Furthermore, in some services, these higher bonuses are not necessarily contingent on performing the duty or serving in a billet requiring the skill. Consequently, reenlistment bonuses can also serve as a type of proficiency pay.

Summary of Findings from Four Military-Related Studies of Credential Pay

We identified four studies that have assessed credential pay in the military context. The first, Mackin et al. (2007), examines the relationship between the payment of Foreign Language Proficiency Pay (FLPP)[1] between 1995 and 2005 and the probability that an eligible enlisted member becomes or remains qualified in a critical foreign language, using data on enlisted personnel in each service. The second study, Dierdorff and Surface (2008), has a similar focus and examines the effect of offering a bonus on foreign language skill acquisition among special operation soldiers in the Army. The third is the study commissioned by the 10th QRMC, Davis and Horowitz (2008), that provided a discussion of the advantages and disadvantages of credential pay, also within the context of acting as an alternative to a TIG pay table to provide incentives for performance. The final study, Hosek and Asch (2002), provided an assessment of skill-pay at the request of the U.S. Air Force.

---

[1]   The special pay for foreign language proficiency changed over time in terms of eligibility, dollar amounts, and even the name of the pay. Prior to 2006, the special pay was called Foreign Language Proficiency Pay (FLPP) and was called the Foreign Language Proficiency Bonus (FLPB) after that. A major difference between FLPP and FLPB is that pay levels for FLPB are considerably higher than FLPP. A history of these pays is provided in Mackin et al. (2007).

Mackin et al. (2007)

Individuals receiving military-sponsored language training are required to have their skills assessed and certified initially following training and then recertified on an annual basis. These individuals can then qualify for FLPP. Using a regression framework, Mackin et al. (2007) estimated the relationship between the payment of FLPP and likelihood an eligible enlisted member is qualified in a foreign language for which FLPP is offered. Mackin et al. found that FLPP expected payments have a positive and statistically significant effect on the probability that the eligible individual is qualified, meaning that higher levels of FLPP payments translate to higher numbers of qualified (proficient) personnel with the estimated effects largest for the Army and the Air Force. For example, in the case of the Army, a 10 percent increase in monthly FLPP payments would be associated with a qualification probability by about 3.3 percent.

Dierdorff and Surface (2008)

Individuals receiving military-sponsored language training are required to have their skills assessed and certified initially following training and then recertified on an annual basis. These individuals can qualify for the Foreign Language Proficiency Bonus (FLPB). Dierdorff and Surface (2008) used five years of data, from 1998 to 2005, on U.S. Army Special Operations Forces soldiers. The data focused on their receipt of the FLPB and their subsequent Defense Language Proficiency Test (DLPT) scores to examine the effect on FLPB on subsequent skill acquisition and maintenance. At the time of the study, FLPB was a monthly amount that ranged from $100 to $200 per month depending on demonstrated skill proficiency.

Dierdorff and Surface found that the FLPB is positively related to individual skill change and maintenance. They also found that the frequency with which skill-based pay is received and the total amount is positively associated with skill development and maintenance.

Davis and Horowitz (2008)

In support of the 10th QRMC, Davis and Horowitz provided a discussion of the potential benefits and drawbacks of credential pay. They also presented an assessment of the concept using criteria developed by the DACMC. Davis and Horowitz defined credential pay as extra money service members can receive every month that they hold a specific credential. The pay is independent of their current billet and the source of the credential. Some of the pays DoD offers are like credential pay, namely flight pay for aviators. That is, aviators receive the pay even if they are not in billets requiring them to fly.

Davis and Horowitz identified three major advantages of credential pay. First, it would increase secondary skills among service members. These are skills that are not the members' primary responsibility but may still be useful to the services. Examples provided by Davis and Horowitz include language proficiency (for those not required to have language proficiency), medical first responders, physical fitness, and process improvement. Second, it would decrease training costs by enabling the military services to leverage civilian training in skills that are not specific to the military. Finally, it would provide greater reliability than incentive pays that are based on billet assignment or profession-based pays. In the case of incentive pays, active members can be ordered to a new billet or even a new profession where the pay is not offered. Because they are less reliable, these pays provide less incentive.

The study discussed issues that the authors felt would need to be addressed when implementing credential pay and their proposed solutions to some of them:

- *Individuals with multiple credentials:* If credentials are nested, members should receive pay only for the highest cost credential. If credentials are complementary, member should receive both pays. If credentials are so dissimilar that member could not use both in the same job either in the military or outside of it, then the service need not pay for both. Member should get paid for the more valuable skill.
- *Reservists:* Reservist eligibility may need to be more restricted than for active duty personnel because reservists cannot be ordered into a new assignment in which the credential can be used. This is because reservists have more discretion in choosing their units. For reservists, it may be necessary to tie some credential pays to member's profession or billet.
- *Syncing with other incentives:* The services would need to make sure members who also receive other special and incentive pays were not overcompensated.
- *Changes to the military rank structure:* The rank structure may need to change if the services require the accession of personnel who are already highly skilled.
- *Oscillating rates:* If credential pay rates are set too high, they could attract more people than needed, so the services would need to lower rates. But this would cause fewer people to earn the credential, leading the services to raise rates again, creating an oscillating pattern that would make planning difficult and causes the services to never have the targeted number of credentialed people.

Davis and Horowitz also considered whether credential pay met the objectives for military compensation developed by the 2006 DACMC. They found that credential pay met the majority of these objectives. They argued that it would support force management by allowing the services to improve retention for personnel in high-demand skills. It would support personnel management flexibility by putting decisions about credential pay into the hands of the services, thereby allowing them to adapt faster to changing circumstances. They argued that it would also support simplification if credential pay replaced some of the pays already used by the services. It would also work for both active and reserve personnel, thereby supporting an integrative personnel management approach to the different components. They also argued that credential pay would be efficient in increasing the services' ability to leverage skills acquitted external to the military. It would also promote efficiency if rates were market-based, so that rates were higher for those with skills in greater demand. Furthermore, if the rates were market-based, it would support the objective that military compensation be consistent with individual choice and volunteerism. Finally, they argued that credential pay would be fair insofar as DoD would honor promises made regarding paying for skill acquisition. To the extent that credential pay would vary across personnel, Davis and Horowitz argued that this is no different from other special and incentive pays, such as selective reenlistment bonuses, that vary across military personnel.

Hosek and Asch (2002)

The Air Force asked RAND to consider ways to strengthen the compensation system for Air Force personnel, focusing specifically on skill pay versus capability pay. Skill pay is pay for designated skills, whereas capability pay is pay based on individual capability, especially current and prospective future leadership potential. To learn about what role these pays might play, Hosek and Asch (2002) reviewed the Air Force's manpower situation, examined data on the level and compensation of military compensation, and considered the advantages and disadvantages of skill pay and capability pay.

Skill pay would emphasize skill. Hosek and Asch argued that the key rationale for skill pay is to protect a valuable stock of current and future human capital when replacing that stock is costly and time-consuming. It would be necessary to define the term *skills* and to establish a program to maintain skills and certify that they have been maintained. Skill pay would help conserve the stock of designated skills that are valuable for military capability and that might be costly and time-consuming to replace. These skills might also be in high demand in the private sector. Compared with bonuses, Hosek and Asch argue that skill pay would have the advantage of being more stable. Bonuses, in contrast, are intended to prevent or address shortages in the flow of personnel currently needed to meet manning requirements in certain specialties. Special pays for aviators and physicians exemplify skill pay from the standpoint of this study.

Hosek and Asch state that skill pay could enable the Air Force to give explicit recognition to the differing external market opportunities available to personnel in various skill areas. It could also provide a means of explicitly rewarding and providing incentives for acquiring and maintaining skills that are essential for military readiness and difficult or costly to replace. Skill pay could be paid to those with a given skill even if they are not using that skill in their current assignment. The rationale for this approach would be that it would enable the Air Force to prevent the loss of critical skills and to maintain a ready inventory of the skills in case of loss of that skill or unexpected demand for it in the future. Though it provides some advantages to the Air Force, Hosek and Asch conclude that skill pay is not designed to be a pay-for-performance incentive.

Discussion

The three studies above predate the authorization for skill-incentive pay and proficiency bonus in 37 U.S.C. section 353. Arguably, the language in the authorization reflects some this prior analysis. For example, skill-incentive pay allows the services to offer a pay for a skill that is not tied to a billet or duty assignment. Consequently, members with the requisite proficiency can receive the pay even if they are not currently performing the duty. Skill-incentive pay or credential pay as discussed in past work can help the services meet the requirements for or to ensure an inventory of personnel with needed skills. It enables the services to pay for expertise that could exist in the civilian sector or be developed in the military by raising pay for marketable skills. It also provides more pay stability to the extent that the pay does not turn on or off as members are rotated in and out of duties requiring the skill.

However, from the standpoint of providing pay-for-performance incentives, credential pay falls short. To the extent that increased skill increases performance, skill pay provides an incentive for greater performance. But credential pay is designed to reward skill and not changes in performance and would not increase or decrease when performance is superior or falls short. Consequently, credential pay would not be a means of replacing a TIG pay table as a mechanism for increasing performance incentives.

Summary

The 13th QRMC requested that RAND assess constructive credit for performance and credential pay as alternatives to a TIG pay table in terms of providing increased performance incentives under the current TIS pay table. This chapter summarized our analysis and findings.

We find that the basic pay profiles for fast promoters under the TIS pay table with constructive credit are similar to those under the TIG pay table. Consequently, the profiles are higher with than without constructive credit. The implication of this analysis is that constructive credit for performance is a policy that can broadly replicate the higher pay found under the TIG pay table. Using the DRM estimates for Army enlisted personnel and officers, our simulations indicate that enlisted and officer retention, average ability, and ability sorting would also improve, but not as much as they would under the TIG pay table. That said, the simulations indicate that constructive credit is less efficient than the TIG pay table, meaning the Army could achieve a given force size and improved performance at less cost with the TIG pay table over the TIS pay table. On the other hand, constructive credit is an improvement over the TIS pay table in terms of efficiency, at least in terms of ability sorting. The implication is that constructive credit for performance would be an improvement over current policy, but not as great an improvement as is predicted to occur under the TIG pay table.

Regarding credential pay, we find that skill-incentive pay and the proficiency bonus authorized under section 353 of Title 37 of the U.S. Code beginning in 2016 is designed to provide higher pay to members with critical skills or career fields. Research on the foreign language proficiency bonus indicates that bonuses are positively associated with greater skill proficiency. Nonetheless, skill-incentive pay credential pay is not designed to be a pay-for-performance program that rewards superior performance and reduces pay for those who fall short. Thus, it would not be an effective substitute to the TIG pay table in terms of increasing performance incentives.

CHAPTER SEVEN

Variation in Time to Promotion and Its Impact on Basic Pay

One of the disadvantages of a TIG pay table discussed by previous commissions and studies, and as summarized in Table 2.1, is the concern that a TIG pay table could result in more inequitable differences in pay to the extent that promotion speed differs across personnel because of factors beyond the control of individual members. These previous studies have argued that if promotion speed varies primarily because of supply and demand factors that cause promotion opportunities to vary across personnel and not because of differences in performance, then a TIG pay table would exacerbate pay differences unrelated to performance. That said, past commissions have also argued that this feature of a TIG pay table also has an advantage. The change in promotion opportunities due to changes in retention self-corrects by creating an offsetting retention and recruiting effect. By magnifying the pay differences associated with promotion, this self-correcting effect is stronger under a TIG pay table.

In this chapter, we consider empirical evidence regarding the role of supply and demand factors in promotion speed, focusing on enlisted personnel, for whom promotion speeds are more apt to vary over time and across personnel. First, we examine the extent of the variation in time to promotion within each service and across entry cohorts within a service, focusing on time to promotion to E-4 and E-5. These promotions are the first competitive ones, since promotions to E-2 and E-3 are nearly automatic if members satisfy TIG and training requirements. That is, these promotions are the first opportunity for promotions to respond to differences in performance. Second, we investigate the extent to which the observed variation in time to promotion affects basic pay trajectories over time, under the TIS versus the TIG pay table.

These first two steps are similar in spirit to the analysis in Chapter Two where we examine basic pay over a career for fast versus slow promoters under the TIS versus the TIG pay table. The difference here is that we show more-detailed results about variations in time to promotion across the services, and then focus in on the implications for basic pay trajectories in the early career versus the entire career. Importantly, these first two steps provide context for the third and fourth steps which are unique to this chapter.

Third, we estimate the extent to which variations in time to promotion are attributable to factors outside the individuals' control, such as supply and demand factors. As we show below, our results suggest that, in general, supply and demand factors explain the largest share of variation in enlisted time to promotion among the covariates tested. Finally, to better understand the extent to which variation in pay is reduced when we account for the variation explained by the supply and demand factors, we redo our calculations of the basic pay trajectories under the TIS versus the TIG pay table. We conclude the chapter with a discussion of the implications of these results.

Data

The analysis in this chapter uses data from DMDC's active duty master and pay files. The active duty master file contains an inventory of all individuals on active duty, and the active duty pay file contains monthly pay and compensation data at the individual service member level. We use data on enlisted service members spanning October 2000 through September 2018 who enter active duty between FY2001-FY2013. Entrants in each FY are called entry cohorts or cohorts in our description below.

The sample of enlisted service members are restricted to those without prior enlisted service and who enter active duty as an E-1.[1] Time to promotion is calculated based on when an enlisted member changes pay grade in the active duty pay file, and presented in months from entry to promotion. Thus, time to promotion in this analysis is measured in terms of months of service until promotion, and not months in a given grade until promotion to the next grade. Unless otherwise noted, time to promotion to E-5 is restricted to the FY 2001–FY 2010 cohorts to ensure that service members are observed for enough years to witness an E-5 promotion.

Variation can be defined in different ways. For the purposes of this analysis, we measure variation in time to promotion as the difference between time to promotion for those in the 10th percentile and 90th percentile of the distribution of time to promotion (i.e., those in the 10th percentile are promoted faster than those in the 90th percentile). By taking the difference between these two percentiles, we can approximate how much time to promotion, and by implication basic pay, differs between those who are promoted the fastest versus those who are promoted the slowest, while excluding outliers (i.e., those with extreme values of time to promotion).

Variation in Time to Promotion Across the Services

To examine variation in time to promotion, we calculated the 10th, 25th, 50th, 75th, and 90th percentiles of time to promotion by cohort for each service. Table 7.1 presents the median time to E-4 and E-5 by service, demonstrating that time to promotion varies across the services. Army enlisted service members are promoted the fastest, with a median time to E-4 of 24 months and a median time to E-5 of 46 months. Marine Corps and Navy enlisted service

Table 7.1
Median Time to Promotion, by Service (months)

Service	Median Time to E-4	Median Time to E-5
Army	24	46
Air Force	30	60
Marine Corps	32	51
Navy	32	52

SOURCE: Authors' calculations.

---

[1] Within each service, we excluded service members in occupations with small sample size, namely, service members with three-digit DoD occupation codes that have 50 or fewer observations in any given cohort were dropped.

members experience the longest median time to E-4, 32 months, while Air Force enlisted service members experience the longest median time to E-5, 60 months.

We also find that the variation in time to promotion can differ by cohort, as shown in Figures 7.1 and 7.2 for time to E-4 and to E-5, respectively. In Figure 7.1, variation in time to E-4, i.e., the vertical distance between the 10th (orange line) and 90th percentile (green line) decreases by cohort among Army enlisted and increases by cohort among Air Force enlisted. In addition, unlike the figures for the Marine Corps and Navy, some of the percentiles of time to E-4 for the Army and Air Force are flat, suggesting that there may be rules in place or standard practices that are common across personnel in all occupations that dictate when these enlisted service members are promoted to E-4.

To empirically investigate this further, we plot Kaplan-Meier survival curves that show the probability of being promoted in a specific month of service, conditional on surviving to that month (e.g., the probability that someone is promoted in month 24 conditional on serving on active duty through month 24).[2] The survival curves in Figure 7.3 indeed show that there

Figure 7.1
Variation in Time to E-4, by Cohort and Service

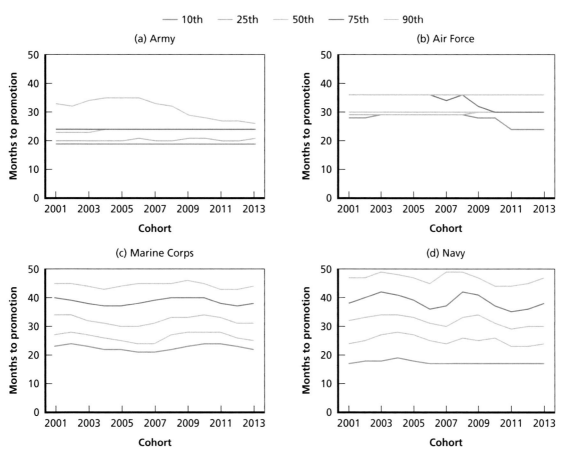

SOURCE: Authors' calculations.

---

[2]   Technically, we calculate the Kaplan Meier survival probabilities to estimate the fraction of service members who "survive" in pay grades E-1 through E-3 before being promoted to E-4, and the figures plot 1 minus these probabilities.

Figure 7.2
Variation in Time to E-5, by Cohort and Service

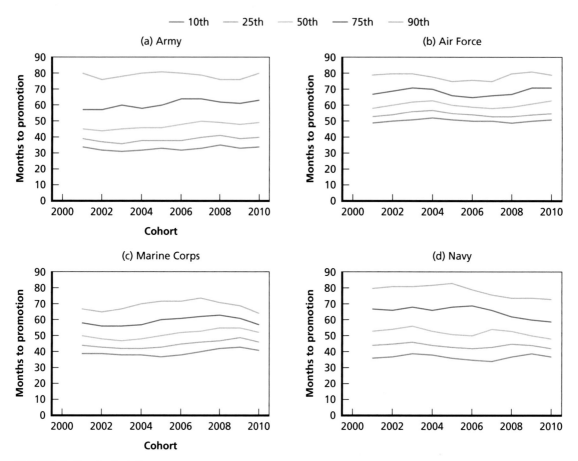

SOURCE: Authors' calculations.

are jumps in the probability of promotion to E-4 at specific months among the Army and Air Force, whereas the curves for the Marine Corps and Navy are generally smooth over time. This suggests that there are certain months when large proportions of enlisted Army and Air Force are promoted. For Army enlisted, there are large increases in the probability of promotion to E-4 between 19 and 24 months. For Air Force enlisted, there are large increases in the probability of promotion to E-4 at 29, 30, and 36 months.

Variation in time to E-4 is roughly constant across cohorts for Marine Corps and Navy enlisted service members, and the variation for these two services is greater than those for Army enlisted and Air Force enlisted. In contrast, Figure 7.2 shows that the Army had the greatest variation in time to E-5, followed by the Navy. Furthermore, variation in time to E-5 was relatively stable across cohorts for both Army and Air Force enlisted service members, with the vertical distance between the 10th and 90th percentiles being roughly the same when comparing time to promotion between the 2001 and 2010 cohorts. For Marine Corps and Navy enlisted service members, variation in time to E-5 decreased for recent cohorts. In particular, for the Marine Corps, variation in time to E-5 decreased between the 2007 and 2010 cohorts, with the difference between the 10th and 90th percentiles reducing from 34 months

Figure 7.3
Kaplan-Meier Survival Curves for the Probability of Being Promoted to E-4, by Service

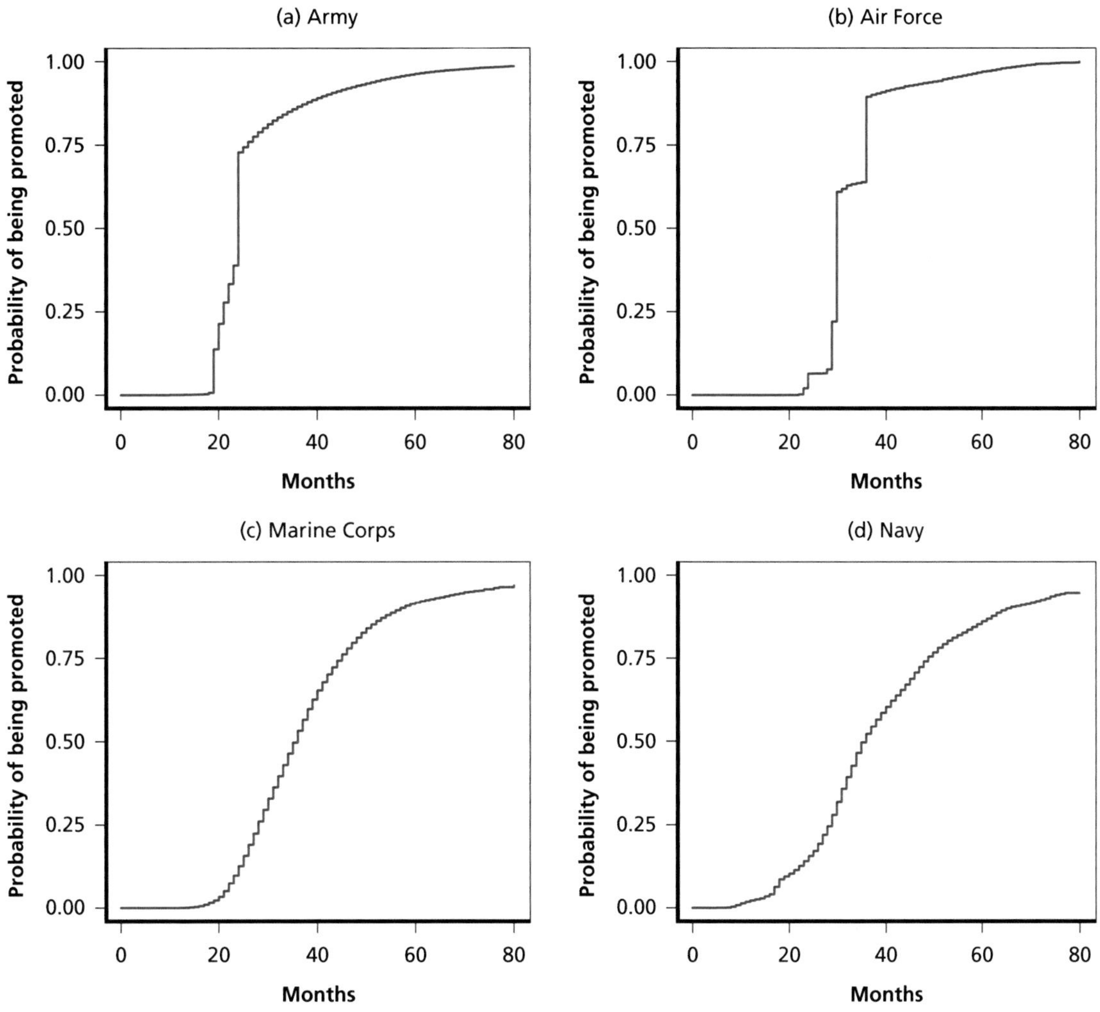

SOURCE: Authors' calculations.

to 23 months. For the Navy, the difference between the 10th and 90th percentiles decreased from 47 months for the 2005 to 36 months for the 2010 cohort.

These results suggest that enlisted time to promotion varies across the services. As we illustrate in the next section, this variation can cause the impact of the TIG pay table on pay trajectories to vary across the services as well.

The Effect of Variation in Time to Promotion on Variation in Basic Pay

To understand how variation in time to promotion impacts pay, we estimate basic pay trajectories under both the current TIS pay table and TIG pay table. We use the January 2020 enlisted pay table (DoD Defense Finance and Accounting Service, 2020) to calculate basic pay under the TIS pay table and use Table 2.2 to calculate basic pay under the TIG pay table.

Because this analysis is meant to be illustrative, we make simplifying assumptions and restrict the sample to service members who are promoted within certain timeframes.

When estimating how variation in time to E-4 affects pay, we restrict the sample to service members with at least 6 YOS who were promoted within 6 years. Thus, individuals with at least 6 YOS who were not promoted within 6 YOS were excluded. We then calculate the 10th and 90th percentile of time to E-4 for each service and estimate pay for the first 6 YOS twice, once assuming that time to E-4 equals to the 10th percentile and once assuming that time to E-4 equals the 90th percentile. The trajectories are calculated assuming the median time to E-2 and median time to E-3 among those in the 10th and 90th percentiles, respectively. Because a large share of these service members were also promoted to E-5 during the first 6 years,[3] we account for E-5 promotions in the basic pay trajectories by estimating a weighted average of pay where the weights are equal to the proportion of service members promoted to E-5 within 6 years and the median time to E-5 for the 10th and 90th percentile of time to E-4 are applied, respectively.

To estimate how variation in time to E-5 affects pay, the sample is restricted to service members with at least 8 YOS who were promoted to E-5 within 8 years. Similar to the E-4 analysis, we then calculate the 10th and 90th percentile of time to E-5 for each service and estimate pay for the first 8 YOS under both the current TIS table and the TIG pay table. The trajectories are estimated using the median times to E-2, E-3, and E-4 for service members in the 10th and 90th percentile of time to E-5.[4]

Table 7.2 shows the 10th and 90th percentiles of time to E-4 and E-5 for the subsamples used for this portion of the analysis. Similar to before, we find that the difference between months to promotion between the 10th and 90th percentiles of time to E-4 are greatest for

Table 7.2
Months to Promotion from Entry for Subsamples

	Army	Air Force	Marine Corps	Navy
Months to E-4				
10th percentile	19	24	21	17
90th percentile	35	36	44	48
Difference	16	12	23	31
Months to E-5				
10th percentile	34	51	41	39
90th percentile	80	79	72	81
Difference	46	28	31	42

SOURCE: Authors' calculations.

---

[3] Among enlisted service members promoted to E-4 within 6 years, 63 percent of Army service members, 73 percent of Air Force service members, 87 percent of Marine Corps service members, and 71 percent of Navy service members were promoted to E-5.

[4] We do not account for promotion to E-6 when calculating the basic pay trajectories for the those in the 10th and 90th percentiles of time to E-5 since a minority of service members are promoted within the first 8 years.

service members in the Marine Corps and Navy, while the difference between the 10th and 90th percentiles of time to E-5 are greatest for service members in the Army and Navy.

Figure 7.4 presents the estimated annual differences in basic pay between those in the 10th percentile and those in the 90th percentile of time to E-4 under both pay tables. Under the TIS pay table, the differences in basic pay between the 10th and 90th percentiles of time to E-4 (blue bars) vary by YOS. In general, those who are promoted faster (i.e., in the 10th percentile of time to E-4) temporarily have greater pay than those who are promoted slower (i.e., in the 90th percentile of time to E-4 under the TIS pay table).

For example, panel (a) of Figure 7.4 shows that the difference in basic pay for Army personnel falls in the 4th and 6th YOS when those in the 90th percentile of time to E-4 are promoted to E-4 and E-5, respectively. A broadly similar pattern is seen for the Marine Corps and Air Force. But for Navy personnel, basic pay for those in the 10th percentile of time to E-4 is greater than pay for those in the 90th percentile for all YOS under the TIS pay table. By the time those Navy personnel in the 90th percentile are promoted to E-4, a large fraction of those in the 10th percentile are promoted to E-5. In other words, Navy enlisted service members who are promoted in the 90th percentile of time to E-4 are not able to catch up pay wise to those who are promoted in the 10th percentile. In contrast to the results for the TIS pay table, basic pay differences between the 10th and 90th percentile of time to E-4 under the TIG pay table

Figure 7.4
Basic Pay Differences Between the 10th and 90th Percentile of Time to E-4, by Service

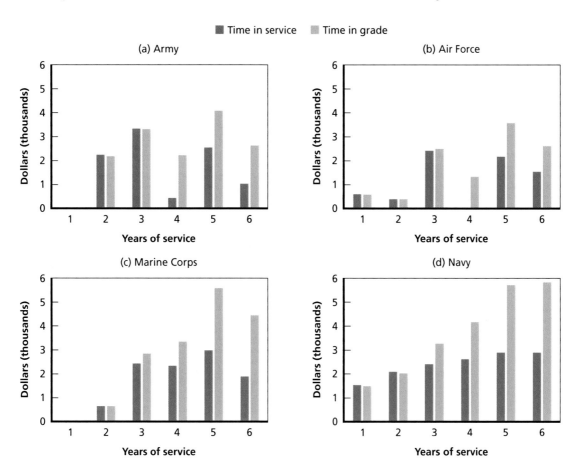

generally remain once those in the 10th percentile are promoted to E-4 (orange bars), or the declines at YOS 4 and 6 are not as great as they are under the TIS pay table. The maximum annual differences across the services are $3,300 (Army, YOS 3) under the TIS pay table and $5,800 (Navy, YOS 6) under the TIG pay table.

The differences in pay gaps across the services between the 10th and 90th percentiles of time to E-4 shown in Figure 7.4 are driven by the differences in variation in time to promotion across the services. Air Force enlisted members experience the smallest variation in pay, and Navy members experience the greatest variation. Taking the total difference across the first 6 years, this amounts to a difference in pay ranging from $7,093 among Air Force enlisted personnel to $14,426 among Navy enlisted under the TIS pay table (Table 7.3). Compared with the TIS pay table, these differences in pay are much larger under the TIG pay table with the TIG pay differences equaling at least 1.5 times of those under the TIS pay table. These results are consistent with our findings in Chapter Two where we find that basic pay is higher over a career under the TIG pay table.

Figure 7.5 shows results similar to those in Figure 7.4, for time to E-5. Similar to the results in Figure 7.4 and the results shown in Chapter Two, the pay differences are larger under the TIG pay table than under the TIS pay table. In addition, the annual differences in basic pay between members in the 10th versus the 90th percentile of time to E-5 are eliminated under the TIS pay table once members in the 90th percentile are promoted to E-5, as shown by the blue bars disappearing at YOS 8 for Army, Air Force, and Navy enlisted and disappearing at YOS 7 for Marine Corps enlisted personnel. The maximum annual difference in pay under the TIS pay table is about $3,000 for Army, Air Force, and Marine Corps enlisted service members and about $3,400 for the Navy. Under the TIG pay table, the maximum annual differences in pay between service members in the 10th and 90th percentile of time to E-5 are larger at about $4,700 for the Army, $3,500 for the Air Force, $5,600 for the Marine Corps, and $6,100 for the Navy.

Table 7.4 shows that the total differences in basic pay across the first 8 YOS vary across the services. In particular, under the TIS table, the total difference in pay between members in the 10th and 90th percentile of time to E-5 varies from $8,091 for the Air Force up to $16,841 for the Navy, with the Navy difference being over twice as large as that for the Air Force.

The variation in total pay differences across the services is even greater under the TIG pay table. They range from $12,077 for the Air Force to $31,928 for the Navy (or over 2.6 times the difference for the Air Force). Moving from the TIS pay table to the TIG pay table would increase the pay difference between members in the 10th and 90th percentile of time to E-5, with the difference ranging from 1.5 to 2.2 times the difference under the TIS pay table. We note that although Army enlisted members have greater variation in time to E-5 than do

Table 7.3
Total Difference over Six Years in Pay Between 10th and 90th Percentile of Time to E-4

Pay Table	Army	Air Force	Marine Corps	Navy
TIS	$9,539	$7,093	$10,239	$14,426
TIG	$14,409	$10,908	$16,802	$22,481
TIG/TIS	1.5	1.5	1.6	1.6

SOURCE: Authors' calculations.

Figure 7.5
Basic Pay Differences Between the 10th and 90th Percentile of Time to E-5

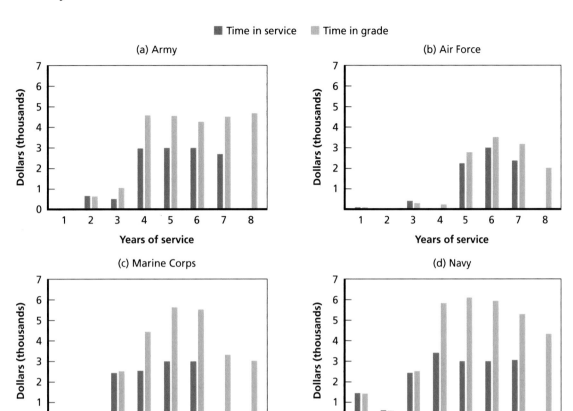

Table 7.4
Total Difference in Pay Between 10th and 90th Percentile of Time to E-5

Pay Table	Army	Air Force	Marine Corps	Navy
Years in service	$12,767	$8,091	$10,909	$16,841
Time in grade	$24,273	$12,077	$24,392	$31,928
Time in grade/years in service	1.9	1.5	2.2	1.9

SOURCE: Authors' calculations.

Navy members, as shown in Figure 7.2, Navy members have a greater variation in the median promotion times to E-2 through E-4 between those in the 10th and 90th percentiles of time to E-5. This explains why differences in pay are larger in Table 7.4 among Navy than Army personnel.

The basic pay trajectories and estimated pay differences between those at the top and bottom of the distributions of time to promotion show that, broadly speaking, greater variation in time to promotion leads to greater variation in pay. Furthermore, because the variation in time to promotion is different across the services, variation in pay is also different across the services. Moving from the TIS pay table to the TIG pay table causes there to be greater variation in pay within a service and across services.

The Extent to Which Variations in Time to Promotion Are Attributable to Observed Factors

Several past commissions rejected the TIG pay table approach because they argued that it would result in inequitable pay differences for members who have different promotion speeds owing to differences in promotion opportunity (supply and demand factors) and not individual merit. In this section, we investigate the extent to which occupation, cohort entry year, or calendar year—that is, factors other than those specific to individuals such as performance—drive variations in time to promotion. We use occupation to capture variation in time to promotion attributable to occupation. As shown in Appendix C, promotion times can vary substantially across occupation within a given service because of different training times and promotion requirements across occupations. Cohort year might explain variation in time to promotion if promotion opportunities varied depending on when individuals enter active duty. Finally, we include the calendar year of promotion to capture factors related to supply and demand conditions at the time of promotion. For example, a robust economy could improve promotion opportunities for those who stay in service, thereby reducing promotion time.

To estimate how much variation in time to promotion is explained by each of these three factors, we use a Cox Proportional Hazards model, a model that accounts for attrition out of the sample if an enlisted member leaves active duty. We then estimate the survival R-squared (Royston, 2006), a measurement of the proportion of variation explained by observable factors similar to the standard R-squared for linear regression models. The model is estimated separately for each service and for each of the three sets of observable factors. We show the R-squared for each of these estimated survival models for time to E-4 and time to E-5 in Tables 7.5 and 7.6, respectively.

In general, calendar year of promotion explains the largest share of variation in time to promotion to E-4 and to E-5, compared with occupation and cohort. In particular, the calendar year promotion dummies in the models explain between 13.1 percent and 47.5 percent of the variation in time to E-4 and between 3.7 percent to 25.1 percent of variation in time to E-5. The one exception is the Navy, where the largest share of variation in time to E-5 is explained by occupation dummy variables (16.8 percent) followed by calendar year promotion dummy variables (14.5 percent). The implication of this analysis is that promotion opportunities in the calendar year of promotion, driven by supply and demand factors, explains more of the variation in promotion times than the other two factors we considered.

To provide more insight on the extent to which variation in time to promotion is explained by when promotions occur, we examine how much variation remains in time to promotion after we account for calendar year of promotion. The remaining variation is of particular interest because it captures factors other than when promotions occur, including individual-specific factors, such as merit.

Our approach involves assessing the extent to which the variation in time to E-4 and time to E-5 is reduced when using predicted times to promotion that account for supply and demand factors as captured by calendar dummies of promotion. For this analysis, we use a negative binomial regression model to estimate the relationship between time to promotion

Table 7.5
Survival R-Squared for Time to E-4

Covariate	Army	Air Force	Marine Corps	Navy
Occupation dummies	0.001	0.007	0.062	0.107
Cohort dummies	0.002	0.004	0.009	0.008
Calendar year promotion dummies	0.284	0.131	0.450	0.475

SOURCE: Authors' calculations.

Table 7.6
Survival R-Squared for Time to E-5

Covariate	Army	Air Force	Marine Corps	Navy
Occupation dummies	0.043	0.004	0.054	0.168
Cohort dummies	0.017	0.009	0.037	0.010
Calendar year promotion dummies	0.141	0.037	0.251	0.145

SOURCE: Authors' calculations.

and calendar year promotion dummies. We then predict time to promotion at the 10th and the 90th percentiles using the regression estimates. If calendar year of promotion fully explains differences in time to promotion, with no remaining differences attributable to other factors, such as individual merit, we would expect no difference between promotion time at the 10th and the 90th percentile. In other words, there would be no remaining variation—everyone would have the same promotion time once we account for supply and demand factors as captured by the calendar dummies. On the other hand, if calendar year of promotion explained only a small amount of the difference in promotion time at the 10th versus the 90th percentile, so that most of the variation were due to other factors, such as individual merit, the remaining portion would be large relative to the observed difference. Thus, we assess the extent of the remaining variation by comparing it to the observed variation. Table 7.7 summarizes our results.

We find that the variation in time to E-4 and time to E-5, as measured by the difference in promotion time at the 10th and 90th percentile, is substantially smaller than the observed differences when using predicted times to promotion that account for calendar year of promotion. For example, the observed difference in time to promotion to E-4 for the Army is 16 months but the differences in predicted times is 4 months. Thus, after accounting for promotion opportunities at the time of promotion, little difference in promotion time is observed. Put differently, other factors including individual merit explain relatively little of the variation in promotion time to E-4 and E-5, regardless of service.

Table 7.7
Variation in Observed and Predicted Months to E-4 and E-5

Percentile of Months to E-4	Army		Air Force		Marine Corps		Navy	
	Observed	Predicted	Observed	Predicted	Observed	Predicted	Observed	Predicted
Months to E-4								
10th percentile	19	22	24	30	21	31	17	31
90th percentile	35	26	36	31	44	33	48	35
Difference	16	4	12	1	23	2	31	4
Months to E-5								
10th percentile	34	50	51	61	41	51	39	55
90th percentile	80	57	79	65	72	59	81	61
Difference	46	7	28	4	31	8	42	6

SOURCE: Authors' calculations.

Implications for Differences in Basic Pay Under the Time-in-Service Versus the Time-in-Grade Pay Table

Tables 7.3 and 7.4 showed observed differences in basic pay between those promoted to E-4 and E-5, respectively, at the 10th versus the 90th percentile under the TIS versus the TIG pay table. In this section, we show the corresponding predicted differences in basic pay, but after controlling for supply and demand factors as measured by calendar year of promotion. The regression models for this analysis are estimated separately for time to E-4 and time to E-5 and separately for each service using the same samples employed to create the earlier basic pay trajectories. This allows us to directly compare our results with those shown in Tables 7.3 and 7.4. As before, we estimate basic pay trajectories both under the TIS and TIG pay table. The results are reported in Tables 7.8 and 7.9 for time to E-4 and to E-5, respectively.

Because there is less variation in predicted time to promotion, the total differences in basic pay between the 10th and 90th percentiles of predicted time to E-4 and predicted time to E-5 are much smaller than their observed counterparts. For example, in Table 7.8, the observed difference in total basic pay for those with promotion times to E-4 at the 10th versus the 90th percentile is $9,539 for the Army under the TIS pay table. This figure reduces to $2,991 when we account for calendar year. Similarly, the figures fall from $14,409 to $4,692 under the TIG pay table. In both cases, the predicted difference is about one-third of the observed difference. For the other services, the predicted difference is even less, as shown in Table 7.8, and as low as 0.05 for the Air Force. We find similar results for time to E-5, shown in Table 7.9.

An implication of the smaller differences in total differences in basic pay using predicted times to promotion is that the amount of the financial incentive for superior performance that remains after accounting for supply and demand factors is relatively small. But the other finding, of particular relevance to the advantages of the TIG pay table, is that the predicted differences are larger under the TIG pay table than the TIS pay table. In other words, if the remaining difference in pay is the incentive for superior performance, that incentive is larger under the TIG pay table, and in some cases, the incentive is considerably larger. For example, in Table 7.9, the pay difference when using the predicted time to E-5 for the Army is almost

Table 7.8
Total Difference in Basic Pay Between the 10th and 90th Percentile of Time to E-4, Observed Versus Predicted

Pay Table	Observed Versus Predicted Promotion	Army	Air Force	Marine Corps	Navy
TIS	Observed	$9,539	$7,093	$10,239	$14,426
TIS	Predicted	$2,991	$383	$2,250	$1,740
TIS: predicted/observed		0.31	0.05	0.22	0.12
TIG	Observed	$14,409	$10,908	$16,802	$22,481
TIG	Predicted	$4,692	$583	$4,103	$1,907
TIG: predicted/observed		0.33	0.05	0.24	0.08

SOURCE: Authors' calculations.

Table 7.9
Total Difference in Basic Pay Between the 10th and 90th Percentile of Time to E-5, Observed Versus Predicted

Pay Table	Description	Army	Air Force	Marine Corps	Navy
TIS	Observed	$12,767	$8,091	$10,909	$16,841
TIS	Predicted	$2,170	$996	$1,992	$2,299
TIS: predicted/observed		0.17	0.12	0.18	0.14
TIG	Observed	$24,273	$12,077	$24,392	$31,928
TIG	Predicted	$5,957	$1,836	$4,805	$4,991
TIG: predicted/observed		0.25	0.15	0.20	0.16

SOURCE: Authors' calculations.

three times higher under the TIG pay table ($5,957) than under the TIS pay table ($2,170). We find broadly similar results, albeit not always as large, for the other services.

Summary

Relative to the TIS pay table, the TIG pay table creates more variation in pay with the extent of variation differing among the services. When testing which observable factors explain variation in time to promotion, we find that promotion opportunities or supply and demand factors, as proxied by calendar year promotion dummy variables, explain the highest share of variation in time to promotion. We predicted time to promotion, accounting for these calendar year effects, and found that variation in time to promotion and, consequently, variation in pay are greatly reduced under both the TIS pay table and the TIG pay table. On the other hand, the remaining variation in pay, the variation that is explained by factors other than calendar year effects, including individual merit, is larger under the TIG than the TIS pay table.

Thus, our results indicate that the conclusions are more nuanced than those drawn by the critics of the TIG pay table. Consistent with the concerns of the critics, we find evidence to indicate that a relatively large share of the variation in promotion is attributable to factors such as supply and demand factors that are unrelated to merit. Further, the TIG pay table would exacerbate the pay differences that result from the variation in promotion. That said, these larger pay differences mean that the TIG pay table would improve the self-correcting retention mechanism that occurs as a result of changes in supply and demand factors, an advantage of a TIG pay table. Furthermore, the remaining differences in pay, representing the financial incentive for performance, are still larger under the TIG than the TIS pay table. This latter finding indicates that the advantage of the TIG pay table, though smaller, still remains, even after accounting for the sizable effects on promotion speed of supply and demand factors.

Discussion and Conclusions

The question motivating this study is whether a TIG pay table would better support the increased focus by the services and Congress on improved talent management and military personnel performance. While interest and consideration of the advantages and disadvantages of a TIG pay table are not new, the 13th QRMC requested that RAND reexamine the merits and drawbacks of a TIG pay table, making use of more-recent data and modeling capabilities such as the DRM. In this chapter. we draw together the findings from the previous chapters to summarize this new evidence regarding the advantages and disadvantages of a TIG pay table using the estimates derived from the specific TIG pay table that we developed for this study. We also review our findings regarding whether the advantages could also be achieved under a TIS pay table with an alteration of current policy and conclude with some final thoughts.

Advantages of The Time-in-Grade Pay Table

The first major advantage of the TIG pay table over the TIS pay table is that the TIG pay table gives a permanent financial reward for early promotion, thereby providing greater incentives for performance, given that fast promotion is the primary means by which the military rewards better performance. Our simulations of basic pay over a career show this to be the case for enlisted personnel and commissioned officers. For example, in simulations of basic pay for enlisted personnel, we find that the discounted present value of basic pay is 11.3 percent rather than 5.5 percent higher for those promoted earlier under the TIG versus the TIS pay table, and the discounted present value of retired pay is 22.8 percent higher, compared with 14.3 percent. Furthermore, the pay advantage of the TIG pay table for those promoted faster remains, even when we control for factors unrelated to performance, such as supply and demand factors that can alter promotion opportunities at a point in time. We find similar results for the other services.

A second advantage is that the TIG pay table provides higher entry pay than the TIS pay table to lateral entrants, thereby increasing the competitiveness of military compensation to individuals with critical civilian-acquired skills, such as cyber skills. These results are consistent with findings of past studies and commissions.

Third, the TIG pay table would be a more efficient approach to setting basic pay. We demonstrate the increased efficiency by making use of the expanded DRM capability we created for this project. The expanded capability allows us to simulate the retention, cost, and performance effects of alternative compensation policies. Because we do not observe ability directly, we parameterize ability so that promotion speed is related to ability, and we calibrate the parameters so that we can replicate the steady-state retention profile of personnel under the TIS pay table. Using the Army enlisted force as an example, we find that greater performance in terms of average ability could be achieved at less cost and for the same retention under the TIG table relative to the TIS pay table. Furthermore, under the TIG pay table, the retention of better performers increases, so the average ability of those in top grades increases relative to ability in the lower grades. Put differently, the Army in our example could achieve the same retention for less cost and achieve a higher-performing force under the TIG pay table.

Finally, if promotions are subject to supply and demand factors, the TIG pay table increases the extent to which promotions help improve retention when these factors change. For example, when the economy improves and retention falls, promotion opportunities improve in occupations that experience the greatest shortfalls. The improved promotion opportunities act as a self-correcting mechanism by inducing higher retention (or lessening the impact of declining retention) and attracting more personnel to occupations experiencing retention issues. Because the TIG magnifies the financial effects of differences in promotion speed, this self-correcting mechanism is stronger under a TIG pay table. As we discuss below in the context of the disadvantages of the TIG pay table, much of the difference in promotion speed is attributable to these supply and demand factors.

Disadvantages of The Time-in-Grade Pay Table

The major disadvantage of the TIG pay table is that the transition would be costly to DoD and would be disruptive to a significant fraction of the force. Examining the YOS and promotion history of active duty personnel in service in January 2019, we estimate that about one-third (32.1 percent) would experience a basic pay reduction in the transition to the TIG pay table, with an average reduction of 6 percent among those who would experience a pay reduction. If DoD adopts a policy to hold members harmless in terms of the level of basic pay by offering save pay, we estimate that in the first year, the cost of this save pay policy would be $1.39 billion in 2018 dollars, with most of the cost attributable to save pay for enlisted personnel. This cost does not include the cost of providing financial education to the force and "socializing" the change to smooth the transition. As discussed by the 10th QRMC and Hogan and Mackin (2008), Congress and DoD could adopt policies to reduce the save pay cost, such as implementing it in conjunction with the annual across-the-board pay rate.

Another challenge with establishing the TIG pay table is the pay for warrant officers and commissioned officers who transition out of the enlisted force could decrease, creating a pay inversion for these personnel. The difficulty is that members promoted from the enlisted force to either the warrant officer or commissioned officer force often have widely different amounts of prior enlisted service. Another difficulty is that the warrant officer TIG pay table is designed for those without prior enlisted service, so pay potentially decreases for those who become warrant officers with prior enlisted service. This disadvantage of the TIG pay table could be addressed by allowing the services to flexibly set the starting grade for those with prior enlisted service. For example, allowing warrant officers with prior enlisted service to transition to warrant officers status at the grade of W-2 or W-3 could address the pay inversion.

Another disadvantage of the TIG pay table is that differences in promotion speed can reflect factors other than differences in individual performance. Although promotion speed is the primary means by which the military rewards better performance financially, promotion speed can differ because of differences in promotion opportunities that arise because of supply and demand factors, as mentioned above. While this self-correcting mechanism is stronger in a TIG pay table, and thus an advantage of the TIG pay table, a critique of the TIG pay table is that the differences in promotion speed would result in more inequitable differences in pay to the extent that promotion speed differs across personnel because of factors beyond their control. Consistent with the concerns of the critics, we find evidence to indicate that a relatively large share of the variation in promotion is attributable to factors such as supply and demand factors that are unrelated to merit. Further, the TIG pay table would exacerbate the pay differences that result from the variation in promotion. But these other factors do not explain all of the differences in promotion speed. To the extent that the remaining differences in pay, after controlling for these other factors, represent the financial incentives for performance, we find that the remaining differences are still larger under the TIG than the TIS pay table. The implication is that while the criticism has merit, it still the case that the TIG pay table provides a stronger financial incentive for performance.

Could the Advantages of the Time-in-Grade Pay Table Be Fully Achieved with a Time-in-Service Pay Table?

The answer to this question is yes for some advantages of the TIG pay table, but in terms of the major advantages of the TIG pay table—the increased efficiency and performance of the force—the answer is no, though, with some changes in policy, a TIS pay table might be able to come close.

As mentioned, an advantage of the TIG pay table is that it would allow pay to be more competitively set for lateral entrants. We find that an identical result could be achieved under a TIS pay table, if Congress changed the current definition of constructive credit to give the services the opportunity to offer not just a higher entry grade but also a higher longevity entry point. For example, a lateral entrant could be permitted to enter as an O-3 with 10 YOS rather than 1 YOS.

We also investigated the further expansion of the definition of constructive credit so that it would give YOS credit in the pay table for better performance. In particular, personnel who are promoted faster than their peers would receive a permanent 1 YOS leg up in the TIS pay table for the purpose of computing basic pay. The purpose of the policy would be to provide a permanent reward for fast promotion, something that is missing under the TIS pay table. Note, however, that this definition would not affect the definition of YOS for the purposes of retirement eligibility or computing retired pay.

We find that constructive credit for performance is a policy that can broadly replicate the higher basic pay found under the TIG pay table. That is, the basic pay profiles for those promoted early under a TIS pay table would be similar to those under the TIG pay table if these individuals received constructive credit for performance. Using the DRM estimates for Army enlisted personnel and officers, our simulations indicate that constructive credit for performance would be an improvement over the TIS pay table by itself in terms of efficiency, at least in terms of ability sorting. But enlisted and officer retention, average ability, and ability sorting would not improve as much as predicted under the TIG pay table. In other words, the simulations indicate that constructive credit is an improvement over the current TIS pay table but would be less efficient than the TIG pay table.

As an additional point, we examined whether credential pay is a policy that could provide performance incentives under a TIS pay table. We find that credential pay is not designed to be a pay-for-performance program that rewards superior performance and reduces pay for those who fall short. Thus, it would not be an effective substitute to the TIG pay table in terms of increasing performance incentives.

Closing Thoughts

Our analysis indicates that the TIG pay table has distinct advantages, especially in terms of supporting service and congressional efforts to improve talent management. But transitioning to the TIG pay table would involve costs, not the least of which is the disruption to the force regarding a fundamental feature of their service—namely how they are paid. While alternative policies, such as constructive credit for performance could achieve some of the advantages of the TIG pay table, simulations suggest that they would not be quite as efficient or performance-enhancing as the TIG pay table. One approach to implementing the TIG pay table while minimizing risk is to do so on an experimental basis. For example, the federal civil service has created "excepted service" for some communities of federal personnel, such as the cyber workforce, and created demonstration projects. In both cases, personnel are paid under a schedule other than the General Schedule. A DoD TIG demonstration project would enable DoD to further assess the retention, cost, and performance effects of the TIG pay table in a "real" setting, as well as gauge the buy-in on the part of the services and members, especially those whose performance is superior, and fully assess the full array of transition costs including the cost of financial education. Should DoD move in this direction, an important first step would be to design such a demonstration project, including the data collection process to ensure rigorous evaluation of the demonstration project.

Time-in-Service Pay Table for January 2018

Table A.1 shows the basic pay table for January 2018.

Table A.1
January 2018 Monthly Basic Pay (Time-in-Service) Table (0–20 Years in Grade)

Grade	Under 2	2	3	4	6	8	10	12	14	16	18	20
Commissioned Officers												
O-10	0.00	0.00	0.00	0.00	0.00	0.00	0.00	0.00	0.00	0.00	0.00	15,800.10
O-9	0.00	0.00	0.00	0.00	0.00	0.00	0.00	0.00	0.00	0.00	0.00	14,696.40
O-8	10,398.60	10,739.40	10,965.60	11,028.60	11,310.90	11,781.90	11,891.40	12,339.00	12,467.40	12,852.90	13,410.90	13,925.10
O-7	8,640.60	9,041.70	9,227.70	9,375.30	9,642.60	9,906.90	10,212.30	10,516.80	10,822.20	11,781.90	12,591.90	12,591.90
O-6	6,552.30	7,198.50	7,671.00	7,671.00	7,700.40	8,030.40	8,073.90	8,073.90	8,532.60	9,343.80	9,819.90	10,295.70
O-5	5,462.40	6,153.60	6,579.00	6,659.40	6,925.50	7,084.20	7,434.00	7,690.80	8,022.30	8,529.60	8,770.50	9,009.30
O-4	4,713.00	5,455.50	5,820.00	5,900.70	6,238.50	6,601.20	7,052.70	7,403.70	7,647.60	7,788.00	7,869.30	7,869.30
O-3	4,143.90	4,697.10	5,069.70	5,527.80	5,793.00	6,083.40	6,271.20	6,580.20	6,741.60	6,741.60	6,741.60	6,741.60
O-2	3,580.50	4,077.90	4,696.20	4,854.90	4,955.10	4,955.10	4,955.10	4,955.10	4,955.10	4,955.10	4,955.10	4,955.10
O-1	3,107.70	3,234.90	3,910.20	3,910.20	3,910.20	3,910.20	3,910.20	3,910.20	3,910.20	3,910.20	3,910.20	3,910.20
Commissioned Officers with over 4 Years of Active Duty Service as an Enlisted Member or Warrant Officer												
O-3E	0.00	0.00	0.00	0.00	5,793.00	6,083.40	6,271.20	6,580.20	6,840.90	6,990.90	7,194.60	7,194.60
O-2E	0.00	0.00	0.00	0.00	4,955.10	5,112.60	5,379.00	5,584.80	5,738.10	5,738.10	5,738.10	5,738.10
O-1E	0.00	0.00	0.00	0.00	4,175.40	4,329.90	4,487.70	4,642.80	4,854.90	4,854.90	4,854.90	4,854.90
Warrant Officers												
W-5	0.00	0.00	0.00	0.00	0.00	0.00	0.00	0.00	0.00	0.00	0.00	7,614.60
W-4	4,282.50	4,606.50	4,738.50	4,868.70	5,092.80	5,314.50	5,539.20	5,876.40	6,172.50	6,454.20	6,684.90	6,909.60
W-3	3,910.80	4,073.70	4,240.80	4,296.00	4,470.60	4,815.30	5,174.10	5,343.30	5,538.90	5,739.90	6,102.30	6,346.80
W-2	3,460.50	3,787.80	3,888.60	3,957.60	4,182.30	4,530.90	4,703.70	4,873.80	5,082.00	5,244.60	5,391.90	5,568.30
W-1	3,037.50	3,364.50	3,452.40	3,638.10	3,857.70	4,181.70	4,332.60	4,543.80	4,751.70	4,915.50	5,065.80	5,248.80
Enlisted Members												
E-9	0.00	0.00	0.00	0.00	0.00	0.00	5,173.80	5,290.80	5,439.00	5,612.40	5,788.20	6,068.70
E-8	0.00	0.00	0.00	0.00	0.00	4,235.40	4,422.60	4,538.70	4,677.30	4,828.20	5,099.70	5,237.40
E-7	2,944.20	3,213.30	3,336.60	3,499.20	3,626.70	3,845.10	3,968.40	4,186.80	4,368.90	4,493.10	4,625.10	4,676.10
E-6	2,546.40	2,802.30	2,925.90	3,046.20	3,171.60	3,453.60	3,563.70	3,776.70	3,841.50	3,888.90	3,944.10	3,944.10
E-5	2,332.80	2,490.00	2,610.30	2,733.30	2,925.30	3,125.70	3,290.70	3,310.50	3,310.50	3,310.50	3,310.50	3,310.50
E-4	2,139.00	2,248.50	2,370.30	2,490.60	2,596.50	2,596.50	2,596.50	2,596.50	2,596.50	2,596.50	2,596.50	2,596.50
E-3	1,931.10	2,052.30	2,176.80	2,176.80	2,176.80	2,176.80	2,176.80	2,176.80	2,176.80	2,176.80	2,176.80	2,176.80
E-2	1,836.30	1,836.30	1,836.30	1,836.30	1,836.30	1,836.30	1,836.30	1,836.30	1,836.30	1,836.30	1,836.30	1,836.30
E-1>4	1,638.30	1,638.30	1,638.30	1,638.30	1,638.30	1,638.30	1,638.30	1,638.30	1,638.30	1,638.30	1,638.30	1,638.30
E-1<4	1,514.70	0.00	0.00	0.00	0.00	0.00	0.00	0.00	0.00	0.00	0.00	0.00

SOURCE: Office of Secretary of Defense, Directorate of Compensation.

Table A.1—continued
January 2018 Monthly Basic Pay (Time-in-Service) Table (22–40 Years in Grade)

					Years in Grade					
Grade	22	24	26	28	30	32	34	36	38	40
Commissioned Officers										
O-10	15,800.10	15,800.10	15,800.10	15,800.10	15,800.10	15,800.10	15,800.10	15,800.10	15,800.10	15,800.10
O-9	14,908.80	15,214.50	15,747.60	15,747.60	15,800.10	15,800.10	15,800.10	15,800.10	15,800.10	15,800.10
O-8	14,268.30	14,268.30	14,268.30	14,268.30	14,625.60	14,625.60	14,991.00	14,991.00	14,991.00	14,991.00
O-7	12,591.90	12,591.90	12,656.40	12,656.40	12,909.60	12,909.60	12,909.60	12,909.60	12,909.60	12,909.60
O-6	10,566.60	10,841.10	11,372.40	11,372.40	11,599.80	11,599.80	11,599.80	11,599.80	11,599.80	11,599.80
O-5	9,280.20	9,280.20	9,280.20	9,280.20	9,280.20	9,280.20	9,280.20	9,280.20	9,280.20	9,280.20
O-4	7,869.30	7,869.30	7,869.30	7,869.30	7,869.30	7,869.30	7,869.30	7,869.30	7,869.30	7,869.30
O-3	6,741.60	6,741.60	6,741.60	6,741.60	6,741.60	6,741.60	6,741.60	6,741.60	6,741.60	6,741.60
O-2	4,955.10	4,955.10	4,955.10	4,955.10	4,955.10	4,955.10	4,955.10	4,955.10	4,955.10	4,955.10
O-1	3,910.20	3,910.20	3,910.20	3,910.20	3,910.20	3,910.20	3,910.20	3,910.20	3,910.20	3,910.20
Commissioned Officers with over 4 Years of Active Duty Service as an Enlisted Member or Warrant Officer										
O-3E	7,194.60	7,194.60	7,194.60	7,194.60	7,194.60	7,194.60	7,194.60	7,194.60	7,194.60	7,194.60
O-2E	5,738.10	5,738.10	5,738.10	5,738.10	5,738.10	5,738.10	5,738.10	5,738.10	5,738.10	5,738.10
O-1E	4,854.90	4,854.90	4,854.90	4,854.90	4,854.90	4,854.90	4,854.90	4,854.90	4,854.90	4,854.90
Warrant Officers										
W-5	8,000.70	8,288.40	8,606.70	8,606.70	9,037.80	9,037.80	9,489.00	9,489.00	9,964.20	9,964.20
W-4	7,239.90	7,511.10	7,820.70	7,820.70	7,976.70	7,976.70	7,976.70	7,976.70	7,976.70	7,976.70
W-3	6,492.90	6,648.30	6,860.10	6,860.10	6,860.10	6,860.10	6,860.10	6,860.10	6,860.10	6,860.10
W-2	5,684.10	5,775.90	5,775.90	5,775.90	5,775.90	5,775.90	5,775.90	5,775.90	5,775.90	5,775.90
W-1	5,248.80	5,248.80	5,248.80	5,248.80	5,248.80	5,248.80	5,248.80	5,248.80	5,248.80	5,248.80
Enlisted Members										
E-9	6,306.60	6,556.20	6,939.00	6,939.00	7,285.50	7,285.50	7,650.00	7,650.00	8,033.10	8,033.10
E-8	5,471.70	5,601.90	5,921.70	5,921.70	6,040.50	6,040.50	6,040.50	6,040.50	6,040.50	6,040.50
E-7	4,848.30	4,940.40	5,291.40	5,291.40	5,291.40	5,291.40	5,291.40	5,291.40	5,291.40	5,291.40
E-6	3,944.10	3,944.10	3,944.10	3,944.10	3,944.10	3,944.10	3,944.10	3,944.10	3,944.10	3,944.10
E-5	3,310.50	3,310.50	3,310.50	3,310.50	3,310.50	3,310.50	3,310.50	3,310.50	3,310.50	3,310.50
E-4	2,596.50	2,596.50	2,596.50	2,596.50	2,596.50	2,596.50	2,596.50	2,596.50	2,596.50	2,596.50
E-3	2,176.80	2,176.80	2,176.80	2,176.80	2,176.80	2,176.80	2,176.80	2,176.80	2,176.80	2,176.80
E-2	1,836.30	1,836.30	1,836.30	1,836.30	1,836.30	1,836.30	1,836.30	1,836.30	1,836.30	1,836.30
E-1>4	1,638.30	1,638.30	1,638.30	1,638.30	1,638.30	1,638.30	1,638.30	1,638.30	1,638.30	1,638.30
E-1<4	0.00	0.00	0.00	0.00	0.00	0.00	0.00	0.00	0.00	0.00

SOURCE: Office of Secretary of Defense, Directorate of Compensation.

Derivation of the First-Order Conditions for Optimal Effort

In this appendix, we go through the details of the derivation of the first-order condition we presented at the end of Chapter Three.

The individual's problem is, given that they are in the active component, to choose a level of effort e_t to maximize their utility:

$$\max_{e_t} V^A(k_t) - Z(e_t).$$

To simplify notation, we define the value function $\bar{V}^A(k_t)$ to be the value of staying in the active component net the disutility of effort, like so:

$$\bar{V}^A(k_t) \equiv V^A(k_t) - Z(e_t).$$

Our goal in the following derivations is to show how effort connects to the expected value of the individual's value function by affecting the probability of promotion. We do this by deriving the first-order condition and then examining the resulting expression.

We can rewrite the maximand by expanding the expression for $\bar{V}^A(k_t)$:

$$\bar{V}^A(k_t) = \gamma^A + W_t^{Ag} + \beta EMax\left(\bar{V}^A(k_{t+1}) + \varepsilon_{t+1}^A, V^L(k_{t+1}) + \varepsilon_{t+1}^L\right) - Z(e_t).$$

Then, taking the derivative with respect to e_t, we get:

$$\frac{\partial \bar{V}^A(k_t)}{\partial e_t} = \beta \frac{\partial EMax\left(\bar{V}^A(k_{t+1}) + \varepsilon_{t+1}^A, V^L(k_{t+1}) + \varepsilon_{t+1}^L\right)}{\partial e_t} - Z'(e_t).$$

This expression can be simplified by observing that we have a closed-form solution for *EMax*(. . .):

$$EMax\left(\bar{V}^A\left(k_{t+1}\right)+\varepsilon^A_{t+1},V^L\left(k_{t+1}\right)+\varepsilon^L_{t+1}\right)=\kappa\ln\left(e^{\frac{\bar{V}^A\left(k_{t+1}\right)}{\kappa}}+\left(e^{\frac{V^R\left(k_{t+1}\right)}{\lambda}}+e^{\frac{V^C\left(k_{t+1}\right)}{\lambda}}\right)^{\frac{\lambda}{\kappa}}\right).$$

Taking the derivatives of both sides, we get:

$$\frac{\partial EMax\left(\bar{V}^A\left(k_{t+1}\right)+\varepsilon^A_{t+1},V^L\left(k_{t+1}\right)+\varepsilon^L_{t+1}\right)}{\partial e_t}=\frac{e^{\frac{\bar{V}^A\left(k_{t+1}\right)}{\kappa}}}{e^{\frac{\bar{V}^A\left(k_{t+1}\right)}{\kappa}}+\left(e^{\frac{V^R\left(k_{t+1}\right)}{\lambda}}+e^{\frac{V^C\left(k_{t+1}\right)}{\lambda}}\right)^{\frac{\lambda}{\kappa}}}\frac{\partial\bar{V}^A(k_{t+1})}{\partial e_t},$$

which further simplifies to:

$$\frac{\partial EMax\left(\bar{V}^A\left(k_{t+1}\right)+\varepsilon^A_{t+1},V^L\left(k_{t+1}\right)+\varepsilon^L_{t+1}\right)}{\partial e_t}=\Pr\left(\bar{V}^S\left(k_{t+1}\right)>V^L\left(k_{t+1}\right)\right)\frac{\partial\bar{V}^A(k_{t+1})}{\partial e_t},$$

where

$$\bar{V}^S(k_{t+1})=\bar{V}^A(k_{t+1})+\varepsilon^A_{t+1}.$$

Substituting back into our original expression, we get the following first-order condition:

$$\frac{\partial\bar{V}^A\left(k_t\right)}{\partial e_t}=\beta\Pr\left(\bar{V}^S\left(k_{t+1}\right)>V^L\left(k_{t+1}\right)\right)\frac{\partial\bar{V}^A\left(k_{t+1}\right)}{\partial e_t}-Z'\left(e_t\right)\equiv 0.$$

We can simplify this expression further by substituting for $\bar{V}^A\left(k_{t+1}\right)$:

$$\bar{V}^A\left(k_{t+1}\right)=p^{g+1}_{t+1}\bar{V}^{A(g+1)}\left(k_{t+1}\right)+\left(1-p^{g+1}_{t+1}\right)\bar{V}^{Ag}\left(k_{t+1}\right).$$

By assumption, the probability of promotion p_{t+1}^{g+1} depends on e_t. Taking the derivative, we get:

$$\frac{\partial \overline{V}^A\left(k_{t+1}\right)}{\partial e_t} = \frac{\partial \overline{V}^A\left(k_{t+1}\right)}{\partial p_{t+1}^{g+1}} \frac{\partial p_{t+1}^{g+1}}{\partial e_t} = \left(\overline{V}^{A(g+1)}\left(k_{t+1}\right) - \overline{V}^{Ag}\left(k_{t+1}\right)\right)\frac{\partial p_{t+1}^{g+1}}{\partial e_t}.$$

Finally, substituting back into the original expression, we get:

$$\frac{\partial \overline{V}^A\left(k_t\right)}{\partial e_t} = \beta \Pr\left(\overline{V}^S\left(k_{t+1}\right) > V^L\left(k_{t+1}\right)\right)\left(\overline{V}^{A(g+1)}\left(k_{t+1}\right) - \overline{V}^{Ag}\left(k_{t+1}\right)\right)\frac{\partial p_{t+1}^{g+1}}{\partial e_t} - Z'\left(e_t\right) \equiv 0.$$

Setting the expression equal to 0 and rearranging terms, we get the first-order condition shown above:

$$\Pr\left(\overline{V}^S\left(k_{t+1}\right) > V^L\left(k_{t+1}\right)\right)\beta\left(\overline{V}^{A(g+1)}\left(k_{t+1}\right) - \overline{V}^{Ag}\left(k_{t+1}\right)\right)\frac{\partial p_{t+1}^{g+1}}{\partial e_t} \equiv Z'\left(e_t\right).$$

Probability of Being Promoted, by Service and Occupation

Figures C.1 and C.2 contain Kaplan-Meier survival curves that show the probability of being promoted to E-4 and E-5, respectively, in a specific month of service, conditional on surviving to that month by service and occupation. The separate survival curves in each subfigure represent different occupations within each service.

Figure C.1
Kaplan-Meier Survival Curves for the Probability of Being Promoted to E-4, by Service and Occupation

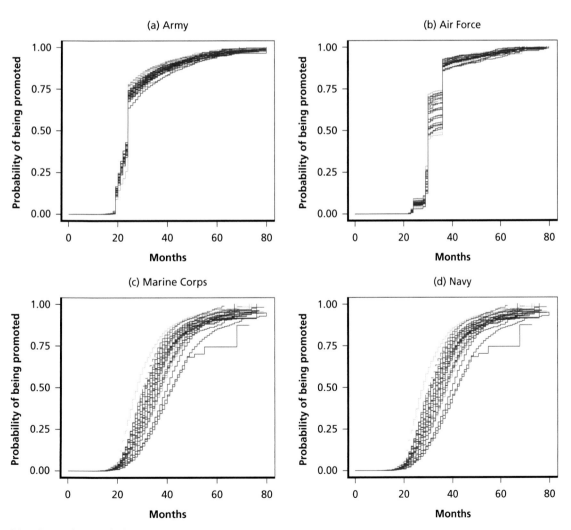

SOURCE: Authors' calculations.

Figure C.2
Kaplan-Meier Survival Curves for the Probability of Being Promoted to E-5, by Service and Occupation

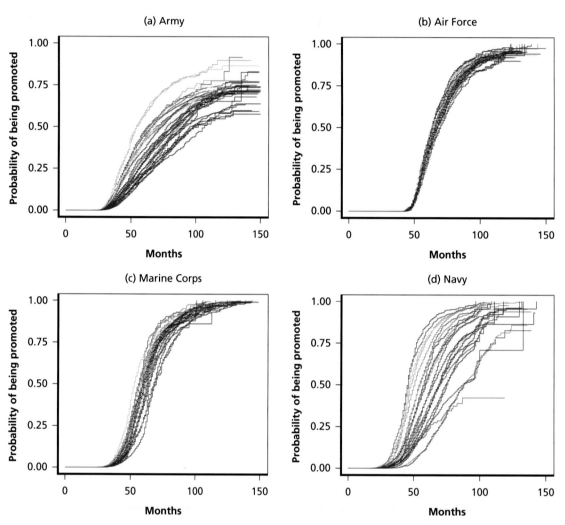

SOURCE: Authors' calculations.

References

Advisory Commission on Service Pay, *Career Compensation for the Uniformed Services: A Report and Recommendation for the Secretary of Defense (the "Hook Commission")*, Washington, D.C.: U.S. Government Printing Office, December 1948.

Asch, Beth J., *Navigating Current and Emerging Army Recruiting Challenges*, Santa Monica, Calif.: RAND Corporation, RR-3107-A, 2019a. As of July 13, 2020:
https://www.rand.org/pubs/research_reports/RR3107.html

Asch, Beth J., *Setting Military Compensation to Support Recruitment, Retention, and Performance*, Santa Monica, Calif.: RAND Corporation, RR-3197-A, 2019b. As of July 13, 2020:
https://www.rand.org/pubs/research_reports/RR3197.html

Asch, Beth J., James Hosek, and Craig Martin, *A Look at Cash Compensation for Active-Duty Military Personnel*, Santa Monica, Calif.: RAND Corporation, MR-1492-OSD, 2002. As of July 13, 2020:
https://www.rand.org/pubs/monograph_reports/MR1492.html

Asch, Beth J., James Hosek, and Michael G. Mattock, *Toward Meaningful Compensation Reform: Research in Support of DoD's Review*, Santa Monica, Calif.: RAND Corporation, RR-501-OSD, 2014. As of April 13, 2020:
http://www.rand.org/pubs/research_reports/RR501.html

Asch, Beth J., James R. Hosek, Michael G. Mattock, and Christina Panis, *Assessing Compensation Reform: Research in Support of the 10th Quadrennial Review of Military Compensation*, Santa Monica, Calif.: RAND Corporation, MG-764-OSD, 2008. As of December 1, 2017:
http://www.rand.org/pubs/monographs/MG764.html

Asch, Beth J., Michael G, Mattock, and James R. Hosek, *The Blended Retirement System: Retention Effects and Continuation Pay Cost Estimates for the Armed Services*, Santa Monica, Calif.: RAND Corporation, RR-1887-OSD, RAND 2017. As of August 20, 2020:
https://www.rand.org/pubs/research_reports/RR1887.html

Asch, Beth J., Michael G. Mattock, James Hosek, and Patricia K. Tong, *Capping Retired Pay for Senior Field Grade Officers: Force Management, Retention, and Cost Effects*, Santa Monica, Calif.: RAND Corporation, RR-2251-OSD, 2018. As of April 2, 2020:
https://www.rand.org/pubs/research_reports/RR2251.html

Asch, Beth J., John A. Romley, and Mark E. Totten, *The Quality of Personnel in the Enlisted Ranks*, Santa Monica, Calif.: RAND Corporation, MG-324-OSD, 2005. As of April 28, 2020:
https://www.rand.org/pubs/monographs/MG324.html

Asch, Beth J., and John T. Warner, *A Theory of Military Compensation and Personnel Policy*, Santa Monica, Calif.: RAND Corporation, MR-439-OSD, 1994a. As of April 28, 2020:
https://www.rand.org/pubs/monograph_reports/MR439.html

Asch, Beth J., and John T. Warner, *A Policy Analysis of Alternative Military Retirement Systems*, Santa Monica, Calif.: RAND Corporation, MR-465-OSD, 1994b. As of August 20, 2020:
https://www.rand.org/pubs/monograph_reports/MR465.html

Asch, Beth J., and John T. Warner, "A Theory of Compensation and Personnel Policy in Hierarchical Organizations with Application to the United States Military," *Journal of Labor Economics*, Vol. 19, No. 3, July 2001, pp. 523–562.

Chu, David, "Reconsidering the Defense Officer Personnel Management Act," testimony to the Subcommittee on Personnel, Committee on Armed Services, Washington, D.C.: U.S. Senate, January 22, 2018. As of March 13, 2020:
https://www.armed-Services.senate.gov/imo/media/doc/Chu_01-24-18.pdf

Congressional Budget Office, *Military Pay and the Rewards for Performance*, Washington, D.C., December 1995. As of March 13, 2020:
https://www.cbo.gov/sites/default/files/104th-congress-1995-1996/reports/doc30-entire.pdf

DACMC—*See* Defense Advisory Committee on Military Compensation.

Davis, Gregory A., and Stanley A. Horowitz, "Military Credential Pay," *The 10th Quadrennial Review of Military Compensation, Volume III, Chapter 8*, Washington, D.C., 2008.

Defense Advisory Committee on Military Compensation, *The Military Compensation System: Completing the Transition to an All-Volunteer Force*, April 2006.

Defense Advisory Committee on Professional and Technical Compensation, *A Modern Concept of Manpower Management and Compensation for Personnel of the Uniformed Services: A Report and Recommendation for the Secretary of Defense, Vol. 1*, Washington, D.C., May 1957.

Dierdorff, Erich C., and Eric A. Surface, "If You Pay for Skills, Will They Learn? Skill Change and Maintenance Under a Skill-Based Pay System," *Journal of Management*, Vol. 34, No. 4, 2008, pp. 721–743.

DoD—*See* U.S. Department of Defense.

Greenbooks—*See* Office of the Under Secretary of Defense for Personnel and Readiness, Directorate of Compensation, *Selected Military Compensation Tables*, 1980–2018.

Gupta, Nina, G. Douglas Jenkins, and William P. Curington, "Paying for Knowledge: Myths and Realities," *National Productivity Review*, Vol. 5, No. 2, Spring 1986, pp. 107–123.

Guthrie, James P., "Alternative Pay Practices and Employee Turnover: An Organizational Economics Perspective," *Group and Organizational Management*, Vol. 25, No. 4, 2000, pp. 419–439.

Hogan, Paul, and Patrick Mackin, "Final Report on the Time-in-Grade Pay Table," *Tenth Quadrennial Review of Military Compensation, Volume III, Chapter 10*, Washington, D.C., 2008.

Hook Commission—*See* Advisory Commission on Service Pay.

Hosek, James, and Beth J. Asch, *Air Force Compensation: Considering Some Options for Change*, Santa Monica, Calif.: RAND Corporation, MR-1566-1-AF, 2002. As of July 13, 2008:
https://www.rand.org/pubs/monograph_reports/MR1566-1.html

Hosek, James, Shanthi Nataraj, Michael G. Mattock, and Beth J. Asch, *The Role of Special and Incentive Pays in Retaining Military Mental Health Care Providers*, Santa Monica, Calif.: RAND Corporation, RR-1425-OSD, 2017. As of July 13, 2020:
https://www.rand.org/pubs/research_reports/RR1425.html

Ledford, Gerald E., and Herbert G. Heneman III, "Skill-Based Pay," Society for Human Resource Management, June 2011. As of January 21, 2020:
https://www.shrm.org/hr-today/trends-and-forecasting/special-reports-and-expert-views/Documents/SIOP%20-%20Skill-Based%20Pay,%20FINAL.pdf

Lockey, Steven, Les Graham, and Qin Zhou, "Evidence for Performance-Related and Skills-Based Pay: Implications for Policing," working paper, Durham University Business School, 2017. As of January 21, 2020:
https://www.npcc.police.uk/Publication/Evidence%20for%20Performance%20Related%20and%20Skills%20Based%20Pay%20Implications%20for%20Policing%202017.pdf

Luthans, Fred, and Marilyn L. Fox, "Update on Skill-Based Pay," *Compensation and Benefit Review*, 1989, pp. 62–67. As of January 21, 2020:
https://journals.sagepub.com/doi/abs/10.1177/088636878902100407

Mackin, Patrick, John Blayne, Paul Hogan, Brian Simonson, and Relja Urgrinic, "Redesigning Foreign Language Proficiency Pay Final Report," SAG Corporation, unpublished manuscript, 2007.

Mattock, Michael, James Hosek, Beth J. Asch, and Rita Karam, *Retaining U.S. Air Force Pilots When the Civilian Demand for Pilots is Growing*, Santa Monica, Calif.: RAND Corporation, RR-1455-AF, 2016. As of December 1, 2017:
http://www.rand.org/pubs/research_reports/RR1455.html

Mitra, Atul, Nina Gupta, and Jason D. Shaw, "A Comparative Examination of Traditional and Skill-Based Pay Plans," *Journal of Managerial Psychology*, Vol. 26, No. 4, 2011, pp. 278–296.

Murray, Brian, and Barry Gerhart, "An Empirical Analysis of a Skill-Based Pay Program and Plant Performance Outcomes," *Academy of Management Journal*, Vol. 41, No. 1, 1998, pp. 68–78.

Office of the Under Secretary of Defense for Personnel and Readiness, Directorate of Compensation, *Selected Military Compensation Tables*, 1980–2018.

Parent, Kevin J., and Caroline L. Weber, "Case Study: Does Paying for Knowledge Pay Off?" *Compensation and Benefits Review*, September–October 1994, pp. 44–50.

Royston, Patrick, "Explained Variation for Survival Models," *Stata Journal*, Vol. 6, No. 1, 2006, pp. 83–96.

Train, Kenneth, *Discrete Choice Methods with Simulation*, 2nd ed., Cambridge, Mass.: Cambridge University Press, 2009.

U.S. Army Recruiting Command, "Army Warrant Officer Recruiting Qualifications," undated. As of May 13, 2020:
https://recruiting.army.mil/ISO/AWOR/BASIC_QUALIFICATION/

U.S. Coast Guard, *Appointing Warrant Officers*, Commandant Instruction Manual M1420.1, Washington, D.C., June 2017.

U.S. Code, Title 37: Pay and Allowances of the Uniformed Services; Chapter 5: Special and Incentive Pays; Section 307: Special pay: special duty assignment pay for enlisted members.

U.S. Code, Title 37: Pay and Allowances of the Uniformed Services; Chapter 5: Special and Incentive Pays; Section 352: Assignment pay or special duty pay.

U.S. Department of Defense, *Defense Modernizing Military Pay, Volume I, Active Duty Compensation, Report of the First Quadrennial Review of Military Compensation*, Washington, D.C.: U.S. Government Printing Office, November 1967.

U.S. Department of Defense, *Report of the Seventh Quadrennial Review of* Military Compensation, Washington, D.C., August 1992.

U.S. Department of Defense, Military Compensation Background Papers, Fifth Edition, Washington, D.C.: U.S. Government Printing Office, September 1996.

U.S. Department of Defense, *Tenth Quadrennial Review of Military Compensation, Volume I, Cash Compensation*, Washington, D.C., February 2008.

U.S. Department of Defense, *Military Compensation Background Papers: Compensation Elements and Related Manpower Cost Items, Their Purposes and Legislative Backgrounds*, 8th Edition, Washington, D.C., 2018. As of April 29, 2020:
https://militarypay.defense.gov/Portals/3/Documents/Reports/Mil-Comp_8thEdition.pdf?ver=2018-09-01-181142-307

U.S. Department of Defense, "Military Personnel Programs (M-1), Department of Defense Budget Fiscal Year 2020," March 2019. As of July 13, 2020:
https://comptroller.defense.gov/Portals/45/Documents/defbudget/fy2020/fy2020_m1.pdf

U.S. Department of Defense, Defense Finance and Accounting Service, "Monthly Rates of Basic Pay (Enlisted), Effective January 1, 2020," 2020. As of January 15, 2020:
https://www.dfas.mil/militarymembers/payentitlements/Pay-Tables/Basic-Pay/EM.html

U.S. Department of Defense, Under Secretary of Defense (Comptroller), *Financial Management Regulation Volume 7A: Military Pay Policy—Active Duty and Reserve Pay*, Washington, D.C., April 2017.

U.S. Government Accountability Office, *Military Compensation: Additional Actions Are Needed to Better Manage Special and Incentive Pay Programs*, Washington, D.C., GAO-17-39, February 2017. As of May 4, 2020:
https://www.gao.gov/products/GAO-17-39

U.S. Marine Corps, *Fiscal Year 2021 (FY21) Enlisted to Warrant Officer (WO) Regular Selection Board*, MARADMIN 682/19, December 10, 2019. As of May 13, 2020:
https://www.marines.mil/News/Messages/Messages-Display/Article/2036993/fiscal-year-2021-fy21-enlisted-to-warrant-officer-wo-regular-selection-board/

U.S. Navy Personnel Command, *The Limited Duty Officer and Chief Warrant Officer Professional Guidebook*, 2011. As of May 13, 2020:
https://www.public.navy.mil/bupers-npc/officer/communitymanagers/active/ldo_cwo/Documents/LDO-CWO_Guidebook.pdf

U.S. Office of Personnel Management, "Fact Sheet: Pay Retention," undated. As of April 29, 2020:
https://www.opm.gov/policy-data-oversight/pay-leave/pay-administration/fact-sheets/pay-retention/

U.S. President's Commission on Military Compensation, *Report of the President's Commission on Military Compensation*, Washington, D.C.: U.S. Government Printing Office, April 1978.